Building and Surveying Series

Accounting and Finance for Building and Surveying   A. R. Jennings
Advanced Valuation   Diane Butler and David Richmond
Applied Valuation, second edition   Diane Butler
Asset Valuation   Michael Rayner
Auctioning Real Property   R. M. Courtenay Lord
Buil ~~~~~~~~~~~~~~~~~~~~~~~~~~ eley
Buil ~~~~~~~~~~~~~~~~~~~~~~~~ rner
Buil ~~~~~~~~~~~~~ 1. Seeley
Buil
Buil ~~~~~~~~~~~~ H. Seeley
Buil
Civi ~~~~~~~~~~ H. Seeley and George
   P. Murray

Commercial Lease Renewals – A Practical Guide   Philip Freedman and Eric
   F. Shapiro
Construction Contract Claim, second edition   Reg Thomas
Construction Economics: An Introduction   Stephen L. Gruneberg
Construction Marketing – Strategies for Success   Richard Pettinger
Construction Planning, Programming and Control   Brian Cooke and Peter
   Williams
Economics and Construction   Andrew J. Cooke
Environmental Science in Building, fourth edition   R. McMullan
Facilities Management, second edition   Alan Park
Greener Buildings: Environmental Impact of Property   Stuart Johnson
Introduction to Building Services, second edition   E. F. Curd and C. A.
   Howard
Introduction to Valuation, third edition   David Richmond
JCT Standard Form of Building Contract 1998 Edition   Richard Fellows
   and Peter Feun
Measurement of Building Services   George P. Murray
Principles of Property Investment and Pricing, second edition   W. D. Fraser
Property Development: Appraisal and Finance   David Isaac
Property Finance   David Isaac
Property Investment   David Isaac
Property Management: a Customer-Focused Approach   Gordon Edington
Property Valuation Techniques   David Isaac and Terry Steley

List continued overleaf

List continued from previous page

Building and Surveying Series
Series Standing Order
ISBN 0–333–71692–2 hardcover
ISBN 0–333–69333–7 paperback
(outside North America only)

You can receive future titles in this series as they are published by
placing a standing order. Please contact your bookseller or, in the
case of difficulty, write to us at the address below with your name
and address, the title of the series and the ISBN quoted above.

Customer Services Department, Macmillan Distribution Ltd
Houndmills, Basingstoke, Hampshire RG21 6XS, England

JCT Standard Form of Building Contract 1998 Edition

A Commentary for Students and Practitioners

Richard Fellows
and
Peter Fenn

Department of Building Engineering
UMIST

palgrave

Parts of this book previously published as
1980 JCT Standard Form of Building Contract
in three editions between 1981 and 1995

First published 2001 by
PALGRAVE
Houndmills, Basingstoke, Hampshire RG21 6XS and
175 Fifth Avenue, New York, N.Y. 10010
Companies and representatives throughout the world

PALGRAVE is the new global academic imprint of
St. Martin's Press LLC Scholarly and Reference Division and
Palgrave Publishers Ltd (formerly Macmillan Press Ltd).

ISBN 0–333–92535–1

This book is printed on paper suitable for recycling and
made from fully managed and sustained forest sources.

A catalogue record for this book is available
from the British Library.

10   9   8   7   6   5   4   3   2   1
10   09   08   07   06   05   04   03   02   01

Printed in Great Britain by
Creative Print & Design (Wales), Ebbw Vale

Contents

Conditions: Part 2: Nominated Sub-Contractors and Nominated Suppliers

Conditions: Part 3: Fluctuations

Conditions: Part 4: Settlement of Disputes – Adjudication – Arbitration – Legal Proceedings

Preface

This book is designed to provide an introduction to the complex contractual situations encountered in the building industry, directly associated with the use of the JCT Standard Form of Building Contract, 1998 Edition.

In 1998 the Joint Contracts Tribunal issued a new set of standard contracts for use on building projects. These documents contain some extensive modifications to the standard contracts (also of JCT origin) which were in wide use. The new contracts not only modified the previous editions but also, by express terms, changed the applicability of some case precedents which had become well known and widely recognised parameters for building operations.

The objective of a standard contract is to provide a clear and unambiguous contract in order to prevent disputes arising out of the terms of the contract and interpretation thereof. Note that provisions in the Unfair Contract Terms Act, 1977, refer to standard form contracts and to consumer contracts; JCT 98 is of the former type.

The legal knowledge required by anyone involved with the use and interpretation of contracts changes and increases daily, so constant attention to the evolution of, particularly, statute and case law is essential.

The intention is to provide an interpretation of and commentary upon the terms of the JCT Standard Form of Building Contract, Private with Quantities, 1998 Edition, combined with considerations of the relevant precedents. Therefore, the aim is to provide a guide for 'everyday' use. In order to pursue any aspect of the contract in great detail, the reader is advised to consult one of the available definitive authorities (such as Hudson's Building and Engineering Contracts).

As this book is intended to serve many purposes, the reader requires no previous knowledge of the standard contracts used in the building industry but will find it of value to have an appreciation of basic English law, particularly that relating to contracts.

The use of this book is recommended to be in conjunction with a copy of the appropriate JCT contract in order that the exact terminology of the document may be studied together with its interpretation. This is particularly important in practical situations where amendments to the contract vary the standard terms.

JCT 1998 is a consolidating document. With one exception it brings all the amendments to the previous contract, JCT 80, within one document.

The exception is the Amendment TC/94 a 'Guide to Terrorism Cover', still printed separately.

This book includes Amendment 1 issued June 1999 and Guidance Notes issued January 2000.

Acknowledgements

Our thanks are due to many people who have been of much help in the preparation and production of this book.

In particular, we are grateful to our academic colleagues at UMIST and in Hong Kong for their support, objective comments and encouragement throughout this venture; Nick Parker deserves special thanks. Also the editorial team at Macmillan (now Palgrave), especially Ester Thackeray, who had the task of ensuring sense and readability of our draft text, for their unwavering patience and care in producing the finished volume.

Our deepest thanks go to our families for their understanding and tolerance throughout our period of working on this book.

Finally, full responsibility for any errors, omissions and contentious statements is ours only.

Peter Fenn, Manchester   *November, 2000*
Richard Fellows, Hong Kong

Table of Cases

Table of Statutes

Abbreviations

AI	Architect's Instruction
BQ	Bill(s) of Quantities
CDP	Contractor's Designed Portion
CITB	Construction of Industry Training Board
CoW	Clerk of Works
DLP	Defects Liability Period
FIDIC	Fédération Internationale des Ingénieurs-Conseils
HGCR	Housing Grants, Construction and Regeneration Act
JCT	Joint Contracts Tribunal
LA	Local Authority
NEDO	National Economic Development Office
NI	National Insurance
NJCC	National Joint Consultative Council
NSC	Nominated Sub-Contract
NS/C	Nominated Sub-Contractor
NSup	Nominated Supplier
PAYE	Pay as You Earn
PC Sum	Prime Cost Sum
PQS	Private Quantity Surveyor
QS	Quantity Surveyor
RIBA	Royal Institute of British Architects
S	Section (in reference to a statute)
S/C	Sub-Contractor
SI	Statutory Instrument
SMM7	Standard Method of Measurement of Building Works: Seventh Edition
SO	Supervising Officer
UK	United Kingdom
VAT	Value Added Tax

Use of this book

This book fulfils three primary functions: as a textbook; as a source to amend and update knowledge of the 1963 and 1980 JCT Standard Forms; and as a practical aid to interpretation and use of the 1998 JCT Standard Form of Building Contract.

It is recommended, therefore, that a copy of the 1998 JCT Standard Form be kept available by the reader to enhance consideration of precise points of detail. In instances of industrial and professional use, reference should be made to a copy of the Contract Document used upon the project under examination (especially where amendments to the standard contract have been incorporated). Reference should also be made to all relevant law reports to ensure compliance with precedents, including recent amendments thereto.

The reader requires no prior knowledge of specialist building contracts, although a basic knowledge of English law, especially contract and tort, and of the building process will prove advantageous.

The reader may occasionally require more detailed discussion than it has been possible to include in this volume; in such circumstances reference is recommended to the works contained in the Bibliography/Sources section to be found at the back of this book.

To accord with the terminology used by the JCT, the style of writing employed in this book uses the masculine gender. In all cases, it is the intention of the authors that masculine versions of words and terms refer equally to their feminine equivalents. No sex discrimination whatsoever is intended and it is hoped that styles of writing will change soon to eliminate such apparent inequity.

Introduction

Under this form of contract, as under its predecessors (RIBA Form, 1963 and 1980 JCT contracts), the contractor undertakes to carry out the work under described conditions. Details of the required work and conditions must be set out in the Contract Documents to be valid.

There are also general principles of law relating to any activity, including construction, which must be observed, e.g. public liabilities, statutes being the primary source of law which must be applied in preference to any other requirements.

Thus the various legislation and Statutory Instruments etc. must always be adhered to even though not specifically incorporated as Contract Documents. They are incorporated into the Conditions of Contract by Clause 6.

The Contract Documents, most particularly the Conditions of Contract, set out the express terms of the contract. Usually, however, further terms must be implied in the contract to give it business efficacy.

Cory v. *City of London Corporation* (1951) 'In general, a term is necessarily implied in any contract, the other terms of which do not repel the implication, that neither party shall prevent the other from performing it.'

Contract Documents (Clauses 1, 2 and 5)
Clause 1.3 specifies the Contract Documents to be:

(a) Contract Drawings
(b) Contract Bills
(c) Articles of Agreement
(d) Conditions
(e) Appendix

However, certain other documents are also incorporated, apart from statute and common law provisions. These are:

(f) SMM7 (Clause 2.2.2.1)
(g) Basic Price List, if applicable (Clause 39.1)
(h) Formula Rules, Monthly Bulletins, if applicable (Clause 40)
(i) Form of Tender.

Form of Tender
Should there be a discrepancy between the Form of Tender and the Conditions of Contract, etc., provided the Contract had been signed by the relevant parties, it is probable that the Contract Conditions would prevail.

1

Specification and Bills of Quantities
The Specification is not a contract document under this form although it
is a Contract Document in the Without Quantities form.

Therefore, the BQ must carry out a multitude of functions, *inter alia*:

(a) an exact measure of the work to be completed for the contract sum
    in terms of both quality and quantity
(b) provide a basis for the measurement and valuation of Variations
(c) provide a means of incorporating the necessary specification infor-
    mation as part of a Contract Document; this is sometimes achieved
    by incorporating the specification as trade preambles in the BQ.

Contractual Relationships
The Parties to the contract are:

The Contractor
The Client (called the Employer)

The Architect, QS, Engineer, CoW, and other consultants, are not
parties to the contract. Each has their own terms of employment with the
Employer (e.g. Architect – Conditions of Engagement), usually of a standard form issued by the appropriate professional institution. Often, such
standard terms set out fee scales – RICS advisory; RIBA advisory.

The Architect is given extensive powers of Agency (of the Employer)
under the JCT Standard Form in respect of the work. He may not alter
the terms of the Contract itself (to which he is not a party).

Clauses 27 and 28 provide for determination of the Contractor's
Employment in certain circumstances. If such determination occurs, it is
the employment of the Contractor under the contract which ceases, prescribed terms of the Contract remain in force to facilitate settlement
between the parties. These clauses are in addition to Common Law rights
where, if one part is in fundamental breach of a contract, the aggrieved
party is relieved of any outstanding obligations under the contract and
usually will have a right to damages also.

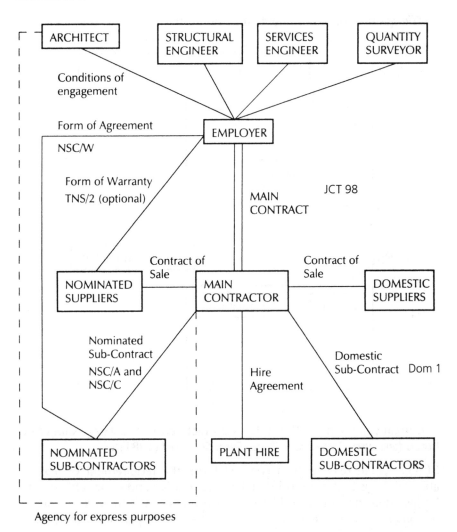

ARCHITECT

STRUCTURAL ENGINEER

SERVICES ENGINEER

QUANTITY SURVEYOR

Conditions of engagement

Form of Agreement

NSC/W

Form of Warranty
TNS/2 (optional)

EMPLOYER

MAIN CONTRACT

JCT 98

Contract of Sale

NOMINATED SUPPLIERS

MAIN CONTRACTOR

Contract of Sale

DOMESTIC SUPPLIERS

Nominated Sub-Contract
NSC/A and NSC/C

Hire Agreement

Domestic Sub-Contract   Dom 1

NOMINATED SUB-CONTRACTORS

PLANT HIRE

DOMESTIC SUB-CONTRACTORS

Agency for express purposes

The nomination components may be further expanded thus:

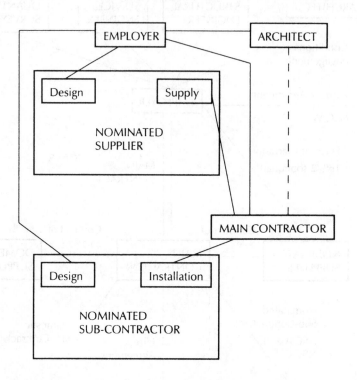

Certain provisions of the Unfair Contract Terms Act, 1977, should be noted (see also: *McCrone* v. *Boots Farm Sales Ltd* (1981)):

S11(1): '... in relation to a contract term, the requirement of reason-ableness ... is that the term shall have been a fair and reasonable one to be included, having regard to the circumstances which were, or ought reasonably to have been, known, or in the contemplation of the parties when the contract was made'.

S17: 'Any term of a contract which is a consumer contract or a standard form contract shall have no effect for the purpose of enabling a party to the contract' to avoid his contractual liabilities unless it is reasonable.

Articles of Agreement

The Articles of Agreement constitute the actual contract between the parties whilst the Conditions of Contract stipulate certain provisions for its execution.

By the Limitation Acts, 1939–80, an action upon a simple contract (executed under hand) must be brought within 6 years of the accrual of the cause of action, 12 years if upon a contract executed as a deed.

Preliminary Article

Sets out certain basics of the contract:

(a)   Date

(b)   Parties – Employer and Contractor

Defines and locates the works	1st Recital
Contractor has given Employer fully priced BQ – 'Contract Bills'	2nd Recital
Numbers the Contract Drawings	3rd Recital
Status of the Employer under Income and Corporation Taxes Act 1988	4th Recital
The extent of the application of the Construction (Design and Management) Regulations 1994 – The CDM Regulations	5th Recital
The Employer has given the Contractor a schedule stating what information the Architect will release and when – Information Release Schedule	6th Recital
If the Employer requires a bond (other than JCT/British Bankers' Association) a copy is provided to the Contractor	7th Recital

Article 1
Overriding obligation on the Contractor to complete the Works, under the described conditions for the Contract Sum.

Article 2
Defines the Contract Sum and refers to the Conditions for the payment times and the payment method.

See Hoenig v. Isaacs (1952)
Lump sum Contract even though there are specific provisions for periodic interim payments.

Note: Housing Grants, Construction and Regeneration Act 1996, Section 109.

Article 3A – Article 3 in Private Edition
Specifies the Architect
   Architect – registered under Architects' Registration Acts, 1931–1969. Architect's death or named Architect ceasing to act as such – Employer must nominate another Architect within 21 days. Contractor has right to object to new Architect within 7 days, subject to arbitration provisions of Article 5. A replacing Architect cannot overrule decisions, instructions, certificates and other similar items issued by previous Architect.
   In *London Borough of Merton v. Stanley Hugh Leach* (1985) the judge accepted that, under JCT 63, the employer was obliged to appoint a person who would be 'reasonably competent and would use that degree of diligence, skill and care in carrying out the duties assigned to him.' Due to similarity of wording, this probably will apply to JCT 80.

Article 3B – LA Form only – not in Private Edition
To be used where the Employer will have a supervising officer instead of an Architect (common for Local Authorities).
   Same provisions as for Article 3A.
   The option of having a Supervising Officer instead of an Architect is a major difference between the Local Authority and Private Forms.

Articles 3A and B
The appointment/re-appointment of Architect, etc. is a condition of the contract – it goes to the root of things.

Note: If the formula fluctuations are applicable under Clause 40, then Clause 40.2 amends the interim valuations provisions of Clause 30.1.2 by requiring an interim valuation to be carried out prior to the issue of each Interim Certificate.
   As the QS is obliged to carry out interim valuations it is submitted that, where Clause 40 is used, the appointment/reappointment of the QS is a condition (rather than a warranty).
   This is evidenced by such provision as the role of the Architect in issuing certificates. Non-appointment/reappointment will be grounds for determination unless the Contract is endorsed to waive this requirement. (*Barnes v. Landair Holdings Limited* (1974).)
   Reappointment must always be of the same status, in the absence of a specific agreement to the contrary.

The SO performs the functions of an Architect under the contract but, due to lack of qualifications and registration is not permitted to be called an Architect.

Article 4
Specifies the QS. Similar reappointment provisions as for Architect.

Note: No specific provision to exclude new QS reversing decisions of previous QS.

Appointment/reappointment – failure to appoint/reappoint – claim for damages for breach of contract.

QS responsibilities under the contract (see also *R.B. Burden* v. *Swansea Corporation* (1957)):

5.1	custody of Contract Drawings and Bills
13.4.1	value Variations under Clause 13.4 rules, unless otherwise agreed by Contractor and Employer
13.6	allow Contractor to be present and take notes etc., for valuing variations
26.1 & 26.4.1	ascertain amount of Contractor's loss/expense – delays – Architect's option make Interim Valuations
30.1.2	make Interim Valuations
30.5.2.1	statement of Retentions – Architect's option
30.6	Final adjustment of Contract Sum
34.3.1	ascertain amount of Contractor's loss/expense – antiquities – Architect's option
38.4.3	agreement with Contractor of amount of fluctuations
39.5.3	agreement with Contractor of amount of fluctuations
40.5	agree with Contractor any alterations of formula fluctuations recovery methods

Note: The Architect, the Supervising Officer and the Quantity Surveyor must carry out their functions under the Contract in the manners expressly provided by the Contract, unless the parties to the Contract agree otherwise.

Article 5 – Dispute or Difference – Adjudication
Provides that the mechanism for settling disputes or differences shall be adjudicated in accordance with Clause 41A.

Article 6
Defines two requirements of the CDM Regulations.

Planning Supervisor – Either the Architect or a Planning Supervisor pursuant to regulation 6(5) of the CDM Regulations.

Principal Contractor – the Contractor or other contractor appointed by the Employer.

Article 7 – Dispute or Difference – Arbitration
Provides that the mechanism for settling disputes or differences under the Contract shall be Arbitration in accordance with Clause 41. Exceptions to this requirement are Clause 31.9 – Statutory Tax Deduction Scheme.

Supplemental Provisions (the VAT Agreement) – clause 3 thereof.

(See also – House of Lords – *Photo Productions* v. *Securicor Transport* (1980). The provisions in the contract govern the outcome and consequences of abandonment or determination.)

Page for the Parties to Sign
The signatories must have authority to sign for this purpose to provide validity to the contract.
    Usually, as specified in the Tender document, the Contract may be executed:

(a)   Under Hand,

    or

(b)   as a Deed.

A1: Agreement executed under hand.
    Employer (or agent) and Contractor (or agent) sign the Contract, each in the presence of a witness (each of whom must sign also).

A2, A3, A4 and A5 concern the Contract being executed as a Deed; the combinations are:
    A2, A3 and A4 – Company (or other body corporate), using (affixing) its common seal, in the presence of its officer(s) (who are authorised to affix the seal),
    A2 and A5 – Company (or other body corporate), names and signatures of either a director and the company secretary or two directors. *Note*: Not to be used by Local Authorities and certain other corporate bodies (exempted under S718(2) of the Companies Act, 1985).
    A2, substitute wording for A3, and A4 – Individual, substitute for A3, 'whose signature is here subscribed' (person must sign), witness to be named and to sign at A4.
    Note: Scottish law requires other attestation clauses to be used.

A6: Executed as a Deed by an individual.

    Limitation Act, 1980 (Statute of Limitations) – actions become 'statute barred' after a specified time lapse from when the cause of action accrued:

(a)   Under Hand – 6 years

(b)   as a Deed – 12 years

In cases involving personal injury, the period of limitation is 3 years from the accrual of the cause of action, which arises when the damage is known or reasonably ought to have been discovered. The situation is more complicated in instances of developing/continuing illness or incapacity, as may occur with some industrial injuries.

Under contract, normally, the cause of action accrues when the contract is broken.

In tort, the accrual of the cause of action has been considered in a variety of cases. The situation following the House of Lords' ruling in *Pirelli General Cable Works* v. *Faber (Oscar) & Partners* (1983) is that the cause of action accrues when relevant and significant damage first occurs, whether it is discoverable or not. An exception may occur where, say, a building is 'doomed from the start' although how and where such a situation applies is unclear; in such a case time begins to run when the building is built.

Exceptions to the principles of time beginning to run occur where fraud is present, such as to conceal damage or a possible cause of future damage. Following *Applegate* v. *Moss* (1971) and s32 of the Limitation Act, 1980, where fraud is present, the limitation period commences at the time when the damage either is discovered or reasonably ought to have been discovered.

See also: *London Borough of Bromley* v. *Rush & Tompkins Ltd* (1985).

In *Billam* v. *Cheltenham B.C.* (1985), it was held that the commencement of the limitation period may be delayed due to ineffective remedying of defects.

Unless the facts show otherwise, there is no contract until the Articles of Agreement have been executed.

William Lacey (Hounslow) Ltd v. *Davies* (1957).

If any work is done prior to this, in expectation of a contract being executed and a contract is not brought into existence, a claim for *quantum meruit* may result. If, however, a contract is executed subsequently, the provision for valuation and payment will apply retrospectively to the work executed. *British Steel Corporation* v. *Cleveland Bridge & Engineering Co. Ltd (1984)*.

Trollope & Colls Ltd v. *Atomic Power Construction Ltd* (1963).
McAlpine Humberoak Ltd v. *McDermott International Inc.* (1992).

However, where the terms of the contract have been (largely) agreed and something like a letter of intent has been recognised as authority for the work to commence, even in the event of a non-execution of a formal contract, a contract will be presumed, the terms being as originally envisaged. This will apply generally (but note responsibilities for work execution, payment, determination, etc.).

Courtney & Fairbairn Ltd v. *Tolaini Bros. (Hotels) Ltd* (1975).

Other relevant cases are:

McCutcheon v. *David MacBrayne Ltd* (1964).
Modern Building (Wales) Ltd v. *Limmer & Trinidad Co. Ltd* (1975).
Brightside Kirkpatrick Eng. Services v. *Mitchell Construction* (1973).
Kitsons Insulation Contractors Ltd v. *Balfour Beatty Building Ltd* (1991).
A. Monk Building & Civil Engineering Ltd v. *Norwich Union Life Insurance Society* (1991).

Note: Winn v. *Bull* (1877) – A document which contains a statement 'subject to contract' will postpone the incidence of liability until a formal document has been drafted and signed, or a contract has been properly entered in an appropriate way.

Sherbrooke v. *Dipple* (1980) – Parties may get rid of the qualification 'subject to contract' only if both expressly agree that the qualification should be expunged or if such agreement must be implied necessarily.

Conditions: Part 1: General

Clause 1: Interpretation, definitions, etc.

1.1   All clause references relate to the Conditions of Contract unless specifically stated otherwise.

1.2   The Contract is to be read as a whole. Thus, any term is subject to qualification by other terms, usually related and cross-referenced, unless specified as to be read in isolation.
        This is a general principle of law.

1.3   Definitions essential to the interpretation of the Contract (most are self-explanatory).
        The definitions listed refer to the Contract as a whole; those given in individual clauses are in the context of that clause only (e.g. Variation – Clause 13.1).
        Following the ruling by the Court of Appeal in *Jarvis John Ltd* v. *Rockdale Housing Association* (1986) the meaning of 'the Contractor' in JCT 80 (and similar contracts) includes the servants and agents of the Contractor (i.e. those through whom the organisation operates). It is probable that the definition does not include sub-contractors.

1.4   Not used.

1.5   Despite the roles and responsibilities of the Architect, and any CoW, the Contractor is wholly responsible for the execution of the Works in accordance with the Contract – Clause 2.1. Only the Final Certificate, as noted in Clause 30.9.1.1, provides conclusive evidence of the Works' compliance with the requirements of the Contract.

1.6   If the Employer appoints or replaces the Planning Supervisor or Principal Contractor he must inform the Contractor in writing.

1.7   If no provision is made for the giving or service of notices or documents they shall be served by an effective means to an agreed address. If no address is agreed then the delivery to the last known principal or registered business address will be treated as effective giving or serving.

1.8   Where the contract requires an act to be done in a specified number of days the period begins immediately; public holidays are excluded.

1.9   The Employer must give written notice to the Contractor that an individual will exercise all the functions ascribed to the Employer's Representative. A footnote says that neither the Architect nor the Quantity Surveyor should be appointed to the role of Employer's Representative.

1.10   The law of the contract shall be English law irrespective of the nationality, residence or domicile of the Parties or the location of the Works.

1.11   Where the Appendix so states, the Parties have agreed to Electronic Data Interchange.

Clause 2: Contractor's obligations

2.1   Primary obligation for the Contractor to execute and complete the Works in accordance with the Contract Documents.

English Industrial Estates Corporation v. *Kier Construction* (1991): (Contract was ICE5.) Following *Yorkshire Water Authority* v. *McAlpine* (1986), the arbitrator held that although a method statement, normally, was not a contract document, because the method statement was attached to the tender (which was a contract document), such attaching caused the method statement to become a contract document.

If the quality is to be the subject of the opinion of the Architect, if must be his reasonable satisfaction (something the BQ should define).

Thus, the Architect cannot demand better quality than is stated in the Contract Documents, as amended by AIs, regarding quality, etc.

By its nature, this condition excludes any design responsibility on the part of the Contractor – he undertakes to complete the Works in accordance with the supplied design; he does not undertake that the building will be suitable for its intended purpose. Neither does the Contractor undertake that the building will not fail, unless due to standards of workmanship, etc., not in accordance with the Contract. He does not contract to produce a result.

See *Cable* v. *Hutcherson Bros. Pty.* (1969) (Australian).

However, the Courts now regard Contractors as experts in construction and so any design matter that the Contractor considers dubious should be queried by him, in writing, to the Architect, requesting his instruction and indicating the possible construction problems and/or possible subsequent failure.

Following *Duncan* v. *Blundell* (1820), *Equitable Debenture Assets Corporation Ltd* v. *Wm. Moss and others* (1984) and *Victoria University of Manchester* v. *Hugh Wilson & Lewis Womersley and others* (1984), an implied term exists for the Contractor to warn the Architect of design defects known to the Contractor and of such defects the Contractor believes to exist. The belief requires more than mere doubt as to the correctness of the design but less than actual knowledge of errors.

Such action should relieve the Contractor of any liability for subsequent building failure due to inadequate or inappropriate design.

'To complete the Works' refers to practical completion, the point at which, *inter alia*, the defects liability period (DLP) commences.

Certain contractual requirements cease at this point also e.g. fire insurance and liquidated damages liability.

'The Architect (or Employer) may dictate the working hours, order of work execution and postponement of work *but not the method of executing the work.*

2.2   .1   Contract Bills may not override or modify provisions of the Articles, Conditions or Appendix but may affect them, e.g. by noting obligations or restrictions imposed by the Employer.

See *Gleeson M.J. Ltd* v. *London Borough of Hillingdon* (1979).

John Mowlem & Co. Ltd v. *British Insulated Callenders Pension Trust Ltd* (1977): (BQ attempted to constrain a performance specification for watertight concrete; consulting engineers used for structural design.) There must be a very clear contractual condition to render a contractor liable for a design fault. Design is a matter which a structural engineer is qualified to execute, which the engineer is paid to undertake and over which the Contractor has no control.

.2.1   Contract bills – to have been prepared in accordance with SMM7. Any departure from SMM7 must be specified in respect of each item or items – usually in the Preambles.

.2   Any error in the BQ or any departure from SMM7 not specified – not to vitiate the Contract but to be corrected and the correction treated as a Variation in accordance with Clause 13.2.

Bryant & Son Ltd v. *Birmingham Hospital Saturday Fund* (1938): If the Bills do not describe the work appropriately, costs of changes necessary are additional to the Contract Sum.

Note: If the error or departure were sufficiently great so as to change the entire basis or nature of the Contract, it is probable that it *would* vitiate the Contract.

See *Pepper* v. *Burland* (1792).

The errors are those concerning BQ preparation by the PQS, not pricing errors by the Contractor.

2.3   Discrepancies in or divergencies between documents.

If the Contractor finds a discrepancy between any of the specified documents – drawing, BQ, AIs (except Variations), drawings or documents issued under Clause 5, and the Numbered Documents (annexed to Agreement NSC/A) – Contractor to give the Architect written notice specifying the discrepancy and the Architect shall issue instructions to solve the problem.

'*If*' indicates that the Contractor is not bound to find any discrepancies but specifies what the Contractor must do should he make such a discovery.

Again, pricing errors by the Contractor are *not* covered nor is his misinterpretation of the design, unless the design is ambiguous, in which case he may claim.

However, the Contractor is considered to be an expert in construction and so the Courts will require him to execute the appropriate duty of care in such things as cross-checking Contracts Documents and drawings.

Thus this situation is by no means crystal-clear; the Contractor should check and only in extreme cases place reliance on this Clause to avoid any liability.

However, in *L.B. Merton* v. *Stanley Hugh Leach Ltd* (1985), the judge accepted that the Contractor was not obliged to check drawings to seek discrepancies or divergencies so as to impose a duty on the Architect to provide further information.

Normally the Contract provisions will prevail over other Contract Documents, e.g. BQ.

See *Gold* v. *Patman & Fotherington* (1958).

Supply of Goods and Services Act (1982): implied term that a supplier, acting in the course of a business, will carry out the service with reasonable skill and care.

2.4   The two parts of this Clause concern Performance Specified Work (as detailed in Part 5 of the Contract) for which the Contractor must provide a statement on how the performance specification will be met; hence, the Contractor must design the work concerned to comply with the specification (fitness for purpose to that extent).

.1   *If* the Contractor finds a discrepancy or divergence between his Statement and AI issued by the Architect after receipt of the Contractor's Statement – Contractor to give the Architect written notice specifying the discrepancy/divergence, Architect to issue an instruction on the matter.

.2   *If* the Contractor or the Architect finds a discrepancy in the Contractor's Statement – Contractor to correct the Statement to remove the discrepancy at no cost to the Employer and inform the Architect in writing of the correction.

Clause 3: Contract Sum – additions or deductions – adjustment – Interim Certificates

Where the Contract provides for the Contract Sum to be adjusted, as soon as any such adjustment amount has been determined, even if partial (or even, presumably, 'on account') it should be included in the computation of the following Interim Certificates.

This is a distinct aid to the Contractor's cash flow, the 'life-blood' of the industry, by requiring even estimate, partial or 'on account' Valuations of extras to be included in the following (Valuation and) certification for payment.

Clause 4: Architect's instructions

4.1.1 Contractor to forthwith comply with all AIs which the Architect expressly empowered by the Conditions to issue.
The exceptions are:

  .1   where such an instruction is one requiring a Variation, the Contractor need not comply if he has made reasonable objection in writing to the Architect

  .2   where such an instruction is one requiring a Variation, the Contractor need not comply if a 13A Quotation has been issued

      (a) until the Architect issues an acceptance of the 13A Quotation

      (b) until an instruction in respect of the Variation has been issued under Clause 13A.4.1.

4.1.2

Receipt by Contractor of written notice from Architect to comply with an AI	Non-compliance – Employer engages another to do the work involved in the AI and contra charges Contractor accordingly

⌊_____⌋

7 days

4.2 Contractor receives what purports to be an AI from the Architect. Contractor may request Architect to specify the authority (which Clause) for the instruction, in writing.
Architect complies with request, forthwith.
Contractor complies with the instruction – deemed, in all cases, to be a valid AI.

4.3.1 All instructions issued by the Architect – to be in writing

  .2 Non-written (verbal) instruction – of no immediate effect. Contractor to confirm the instruction to the Architect within 7 days. Architect has 7 days from receipt of confirmation to dissent in writing, if not becomes valid AI.

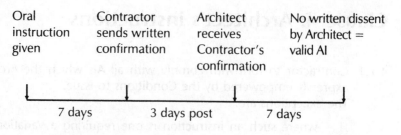

Oral instruction given	Contractor sends written confirmation	Architect receives Contractor's confirmation	No written dissent by Architect = valid AI

| 7 days | 3 days post | 7 days |

.1   Architect may confirm an oral instruction within 7 days – no need for Contractor to confirm – is valid AI from the date of the Architect's written confirmation.

.2   No confirmation of an oral instruction but the Contractor complies – Architect *may* confirm in writing prior to the issue of the Final Certificate – deemed to be a valid AI on the date when given orally.
      This course of action imposes considerable risk of loss upon the Contractor and much reliance on the Architect's memory!

AI

| Must be in writing | If requested Architect states authority in writing (Clause –). If Contractor complies, always deemed to be valid AI | Architect gives Contractor written notice to comply with an AI, Contractor has 7 days to comply. If not, another Contractor and Contras, etc. | Oral instruction. Contractor 7 days to confirm. Architect 7 days to dissent from receipt, if not = valid AI.
■ Architect may confirm oral instruction within 7 days of giving.
■ Contractor complies with oral instruction. Architect may confirm prior to issue of Final Certificate = valid AI. |

Clause 5: Contract documents – other documents – issue of certificates

5.1   Contract Drawings and Bills to remain in the custody of the Architect or QS and be available at all reasonable times (usually normal working hours) for the inspection of the Contractor and Employer.

5.2   Immediately the Contract is executed, the Architect to provide free to the Contractor:

    .1   1 copy of the Contract Documents certified on behalf of the Employer.

    .2   2 further copies of the Contract Drawings.

    .3   2 copies of the unpriced BQ.

5.3.1 As soon as possible after the execution of the Contract:

    .1   Architect to provide free to the Contractor, 2 copies of all descriptive schedules, etc., necessary for use in carrying out the Works.

    .2   Contractor to provide free to the Architect, 2 copies of the master programme. Amendments made within 14 days of Extension of Time award (clause 25.3.1), i.e. to be kept updated in accordance with Architect's Extension or Time awards. (Useful to show when information will be required.) Or from the date of a confirmed acceptance of a 13A Quotation.
          Note: A network will show the effects of delays far more clearly than a bar chart.

    The following points should also be noted:
        (a) There is no definitive statement indicating the form of the programme or what it is to show. The type of programme and information shown is, thus, at the option of the Contractor (in the absence of any agreement denoting specific type of programme and information required). Thus a simple bar chart, a network or any other type of chart will be compliant. Normally, it is suggested that the programme should be of a type in common usage, clearly denoting key dates and work sequences, etc. BQ may note form of programme required.
        (b) If a Contractor disagrees with an extension or other Completion Date modification by the Architect, in order to further his claim, it would be perhaps prudent not to amend the programme

to comply with the Architect's award but to revise the programme to comply with the Contractor's own estimate of the appropriate Extension of Time and to indicate by this means the procedure for completion by that date.

Indeed, it should be borne in mind that there is no provision in the Contract for the Contractor to amend the programme to take account of the Architect's reduction of an Extension of Time (due to work being omitted).

.2 Nothing in the supplementary documents (schedules, master programme, etc.) may impose any obligation beyond those imposed by the Contract Documents.

5.4 Architect to provide free to the Contractor, 2 copies of all necessary drawings and details, supplementary to the Contract Drawings, to enable the Contractor to execute the Works.

5.5 Contractor to keep on site and available for the Architect (or representative) at all reasonable times:

    (a) 1 copy of Contract Drawings
    (b) 1 copy of unpriced BQ
    (c) 1 copy of Schedules, etc.
    (d) 1 copy of master programme (if applicable)
    (e) 1 copy of drawings and details

Reasonable times likely to be construed as site working hours (or normal working hours, if longer).

5.6 Upon final payment, the Contractor to return to the Architect all drawings, schedules, etc., which bear Architect's name, if requested to do so.

5.7 Limits to use of documents.
Documents to be used for the purposes of the Contract only and for no other projects.

Employer, Architect and QS may use the BQ rates or prices only for the purposes of the Contract – also may not divulge them except for the purposes of the Contract.

5.8 Normally, all Certificates must be issued to the Employer by the Architect who must immediately send a duplicate copy to the Contractor.

Exception:
35.15.1 failure of NS/C to complete by appropriate date.

Following *L. B. Merton* v. *Stanley Hugh Leach Ltd* (1985) it should be noted that the following terms are implied:

(a) the Employer will not hinder or prevent the Contractor carrying out his obligations under the Contract or from executing the Works in an orderly and regular way,

(b) the Employer undertakes that the Architect will do all required to enable the Contractor to carry out the work,

(c) where the Architect is required to supply the Contractor with drawings, etc. during the course of the work, those drawings, etc. must be accurate.

5.9 Contractor to provide free to the Employer, drawings and other information describing Performance Specified Work as built and maintenance as specified in the BQ or AI(s) for expenditure of the appropriate provisional sum.

Clause 6: Statutory obligations, notices, fees and charges

6.1   .1   Overriding obligation to comply with any and all Statutory Requirements, including the regulations of statutory undertakers.

       .2   *If* the Contractor finds any divergence between the Statutory Requirements and the Contract Documents, Drawings, etc., or AIs he must immediately inform the Architect, in writing, specifying the divergence.

          Note: the role of 'if', not 'when', and the implications as discussed previously.

       .3   If the Contractor gives notice or by some other means the Architect discovers such divergence, he has 7 days from receipt of any notice in which to issue instructions about the matter. If such instructions require the Works to be varied, this constitutes a valid Variation (Clause 13.2).

       .4   .1, .2, .3, Emergency compliance with Statutory Requirements prior to instructions being received:

          Contractor to supply necessary materials and to execute necessary emergency work.

          Contractor to inform the Architect immediately of the situation and the steps being taken.

          Such materials and work are then deemed to be pursuant to an AI, provided that the necessity for the emergency work was due to a divergence as described – valuation will be made accordingly (Clause 13.2).

       .5   Provided that the Contractor has complied with the requirements regarding divergence, he is not liable where the Works do not comply with any Statutory Requirements, provided that they do comply with the Contract Documents, including Drawings.

          Note: This liability avoidance is contractual only, i.e. expressly restricted to liability to the Employer.

       .6   If the Contractor or Architect finds a discrepancy between the Statutory Requirements and any Contractor's Statement:

          immediately give the other written notice specifying the discrepancy,

Contractor to inform the Architect of the amendment proposed to remove the divergence, and Architect to issue appropriate instructions.

Contractor's compliance with AIs issued under this Clause to be at no cost to the Employer (except as provided in Clause 6.1.7).

.7   Any changes in Statutory Requirements after the Base Date which necessitate a change to Performance Specified Work to be treated as an AI requiring a Variation (Clause 13.2) and valued accordingly.

No Contractor may construct a building which does not comply with the applicable Statutory Requirements.

6.2   Fees and charges – Contractor to pay and indemnify the Employer in respect of all statutory fees and charges.

Such fees and charges, including all taxes except VAT, are added to the Contract Sum unless they:

.1   arise due to a statutory undertaker being a NSup or NS/C; or

.2   are priced in the Contract Bills; or

.3   are a provisional sum in the Contract Bills.

6.3   The contractual provisions regarding assignment and sub-letting (Clause 19) and NS/Cs (Clause 35) do not apply to statutory undertakers carrying out their statutory obligations. For this purpose they are not Sub-Contractors within the meaning of the Contract.

Clause 6A: Provisions for use where the Appendix states that all the CDM Regulations apply

6A.1 The Employer shall ensure:

that the Planning Supervisor carries out all the duties of a planning supervisor under the CDM Regulations

where the Contractor is not the Principal Contractor, that the Principal Contractor carries out all the duties of a principal contractor under the CDM Regulations.

6A.2 Where the Contractor is a Principal Contractor he must comply with all the duties set out in the CDM Regulations, in particular the Health and Safety Plan.

6A.3 Where the Employer appoints another Principal Contractor the Contractor shall comply, at no extra cost, with all reasonable requests and no extension of time shall be given.

6A.4 The Contractor shall provide, and ensure that any sub-contractor provides, any information required for the health and safety file required by the CDM Regulations.

Clause 7: Levels and setting out of the Works

The Architect is obliged to determine and provide the Contractor with all requisite levels and dimensions for the execution of the Works.

The Contractor is responsible for correcting at his own cost any setting out errors for which he is responsible, i.e. his inaccuracies in setting out.

Provided that the Employer has consented, the Architect may issue an instruction permitting an inaccurately set out building to be completed with an appropriate price reduction by the Contractor.

Trespass: If setting out causes a trespass on adjoining property, the injured party has an action:

(a)   against the Contractor if the trespass is caused by setting out error(s) on his part,

(b)   against the Employer (who may be able to recover from the Architect) if the trespass is caused by inaccurate drawings or other information supplied by the Architect.

See *Kirkby* v. *Chessum & Sons Ltd* (1914).

Note: The Contractor may be required to insure against the risks involved:

(a)   Clause 20.2

(b)   Clause 21.2.1.

Clause 8: Work, materials and goods

8.1   Bills to describe all materials, goods and workmanship, thus the
       BQ acts as the specification document. For Performance Specified
       Work, the materials and goods must accord with the Contractor's
       Statement.

       'So far as procurable' modifies this overall requirement – may be
       best to give alternatives, if applicable – e.g. paint manufacturer's list,
       in the Preambles. If this is not done, the phrase may be deemed to
       mean the nearest substitute to the specified item, if unavailable.

       Materials and goods must be to the Architect's reasonable
       satisfaction, in accordance with Clause 2.1.

       This clause has led to the wide use of the term 'or equal and
       approved' – this means the Architect *may* approve other goods,
       not *must*.

       See *Leedsford Ltd* v. *Bradford Corporation* (1956).

       Where described therein, workmanship must be to the standards
       in the BQ or Contractor's Statement; otherwise, workmanship must
       be to a standard appropriate for the Works. Workmanship must be
       to the Architect's reasonable satisfaction, in accordance with Clause
       2.1.

       .3   All work shall be carried out in a proper and workmanlike
             manner and in accordance with the Health and Safety
             Plan. (Following: Court of Appeal – *Greater Nottingham Co-
             Operative Society Ltd* v *Cementation Piling and Foundations Ltd*
             (1988).)

       .4   Contractor must have the Architect's written consent (not to
             be unreasonably withheld or delayed) to substitute any mate-
             rials or goods for those described in any Contractor's State-
             ment. The Contractor retains all responsibilities for any
             properly substituted materials or goods.

8.2.1 Architect may require the Contractor to provide vouchers to prove
       that the goods and materials comply with the specified require-
       ments as Clause 8.1.

       .2   If the Architect is not satisfied with materials, goods or workman-
             ship included in work which has been executed and which is
             required to be to his reasonable satisfaction (in accordance with
             Clause 2.1), the Architect must express such dissatisfaction within
             a reasonable time from the execution of the (alleged) unsatisfac-
             tory work.

Two issues arise – how is the Architect's dissatisfaction to be expressed (and to whom) and how is the Architect to know when each item of work is executed in order to be able to judge the reasonable time period? Sensibly, the Architect should express any dissatisfaction in writing to the Contractor (with copies to any other relevant participants – especially the Quantity Surveyor). The Architect's knowledge of progress of the Works, through monitoring procedures, should indicate the (approximate) time at which the (alleged) unsatisfactory work was executed.

8.3 AI may be issued for the Contractor to open up work for inspection or to test goods and materials.

Costs to be added to the Contract Sum or may be provided in the BQ (Provisional Sum) – including costs of making good.

However, if the tests, etc., prove the work/materials to be defective, the Contractor must bear the costs.

8.4 The four sub-clauses of Clause 8.4 give the Architect authority to order action to be taken by the Contractor if any work, materials or goods are not in accordance with the Contract (i.e. are found not to comply with the specification). The authority is additional to, and does not affect (is without prejudice to), the general powers of the Architect.

.1 Architect may issue an AI for removal from the site of work, goods or materials not in accordance with the Contract requirements.

Note: (a) The Architect should give the reason for the AI, as removal is a Variation under Clause 13.1.1.3 where the goods, etc., *are* in accordance with the Contract.

(b) By Clause 27.2.1.3, the employment of the Contractor may be determined if the defective items are not removed, provided that the items are material to the Works.

(c) Defects becoming apparent due to, *inter alia*, Clause 17.2 unspecified materials, etc., must be made good at the Contractor's expense.

Note: Following *Holland Hannen and Cubitts (Northern) Ltd* v. *Welsh Health Technical Services Organisation* (1985), an Architect must take care over instructions issued regarding apparently defective work. The authority given the Contract Clause 8.4, is for the Architect to instruct the Contractor to remove the defective work, etc.; the Contractor remains under an obligation to complete the work in accordance with the Contract.

.2   Architect may, given agreement of the Employer and after consultation with the Contractor (who must consult any NS/Cs affected), allow work, materials or goods which are not in accordance with the Contract to remain as part of the Works. The Architect must confirm such permission in writing to the Contractor but the permission will:

(a) not constitute a Variation and

(b) command an appropriate deduction against the Contract Sum.

.3   Architect may, after consultation with the Contractor (who must consult any NS/C affected), issue AIs requiring Variations which are reasonably necessary due to instructions to remove defective items from site (Clause 8.4.1) or allowing defective items to remain part of the Works (Clause 8.4.2). Such AIs:

(a) cannot provide grounds for addition to the Contract Sum and

(b) cannot provide grounds for an extension of time.

.4   Architect may, given due regard to the Code of Practice following Clause 42, issue AIs to open up work for inspection/test in order to establish the likelihood/extent of any further examples of work, materials or goods which do not comply with the Contract. The scope of such opening up must be reasonable having regard to the Code of Practice and the criterion of the Architect's reasonable satisfaction.

Irrespective of the results of the inspections/tests, no addition to the Contract Sum may be made (under Clauses 8.3 and 26). However, if the inspections/tests showed that the work, materials and goods did comply with the Contract, the work of opening up, inspections/tests and making good constitutes a Relevant Event for which the Architect must award an extension of time under Clause 25 (NB Clause 25.4.2).

8.5 If any work is not carried out in a proper and workmanlike manner, as required by Clause 8.1.3, the Architect, after consultation with the Contractor (who must consult any NS/C affected), may issue AIs as necessary (which may require a Variation). Such AIs:

(a) cannot provide grounds for addition to the Contract Sum and

(b) cannot provide grounds for an extension of time.

8.6 Architect may (not unreasonably or vexatiously) issue an AI for the dismissal *from the Works* of any person employed thereon. This means removal from that particular project, *not* dismissal from employment.

A person is defined in this context as 'an individual or firm (including bodies corporate)' so firms and individuals may be dismissed.

Note: (a) Interim payments include only items of work properly executed in accordance with the Contract requirements – as this decision of compliance is that of the Architect, in theory the QS may not deduct from his recommendation for payment unless an AI as to unsuitability exists.

In such circumstances, however, the QS may be held not to be fulfilling his role as a professional (expert) and so should include a note of his valuation of any 'suspected defective work'. It is then for the Architect to deduct any appropriate amounts at certification.

See *Sutcliffe* v *Thackrah* (1974).

(b)   From several House of Lords decisions, the Contractor gives warranties or implied terms:

(i)   that the workmanship will be of a good standard,

(ii)   that materials and goods supplied will be free from latent and patent defects (including any under a nomination), and that they

(iii) will be suitable for the purposes for which they are supplied.

These may be expressly excluded, or excluded by the circumstances of the Contract. The Standard Form does *not* refer to these matters and so, unless other Contract Documents give specific reference, they will be implied terms.

See *Young & Marten Ltd* v. *McManus Childs Ltd* (1969) and *Gloucestershire C.C.* v. *Richardson* (1969).

Unfair Contract Terms Act, 1977 – in reference to exemption clauses – Contractor should ensure that the above warranties are passed on to suppliers – note that if the Architect insists on nominating a supplier who will not accept these warranties, the Contractor has no liability in respect of the items in question except for patent defects (defects obvious on reasonable examination).

Limits application of exemption clauses in regard to consumer transactions/sales, including trading on one party's standard conditions – s3. However, contracts let by tender do not constitute consumer sales.

Clause 9: Royalties and patent rights

9.1 All sums payable in respect of patented items are deemed to have been included in the Contract Sum by the Contractor where described by or referred to in the Contract Bills.

Contractor to indemnify the Employer against any claim for infringement of patents by the Contractor.

9.2 Where Contractor uses patented items due to his compliance with an AI, he is not liable for any patent infringements.

Any royalties, damages, etc., so arising to be paid by the Contract must be added to the Contract Sum.

Note: (a) Where patents are infringed by the Contractor and the articles in question are not referred to in the Contract Bills, the Employer may incur liability, depending on the circumstances, provided that the articles were for proper inclusion in the Works.

(b) *Prima facie*, the Employer is not liable in respect of AI provision unless the AI constitutes a Variation.

(c) For the implementation of this clause, nominated items are presumed to be the responsibility of the Architect – the Contractor will not usually be fully conversant with the contents of the nomination at the time of tender – *note*, however, SMM7.

Clause 10: Person-in-charge

Contractor to keep a *competent* person-in-charge on the site (usually taken as normal working hours or actual site working hours). This person is given the power of Agency on behalf of the Contractor in respect of receipt of AIs and CoW directions – hence the common title 'Site Agent'.

All relevant persons should be informed who the Site Agent is and who is the deputy (the person who will assume the responsibilities in the absence of the Agent). Any changes in these personnel should be notified immediately.

Clause 11: Access for Architect to the Works

Access is to be for the Architect and his representatives and at all reasonable times. This applies to the site and all workshops, etc., including those of Sub-Contractors, where items for the project are being made, so far as procurable by the Contractor.

This will, of course, include the CoW, engineers and QS (especially for valuation purposes).

Such access may be subject to reasonable restrictions of the Contractor or any sub-contractor in order that their proprietary rights ('trade secrets', etc.) are protected.

Clause 12: Clerk of works

Employer is entitled to appoint a CoW.

CoW acts solely as an inspector on behalf of the Employer but under the directions of the Architect.

Contractor to give adequate facilities for the CoW to perform his duties.
CoW directions:

(a) given to the Contractor,

(b) no immediate effect,

(c) to be valid must be in respect of a matter about which the Conditions expressly empower the Architect to issue instructions,

(d) confirmed in writing by the Architect within 2 working days of being given to be of any effect – if so confirmed, then from date of confirmation – deemed to be an AI.

Contractor has no contractual power to object to a CoW.

Provided that the Architect has properly briefed the CoW, the Employer is responsible for him.

Leicester Board of Guardians v. *Trollope* (1911) shows that the Architect is responsible for his design being followed – not avoided by a CoW being on site.

Clause 13: Variations and provisional sums

13.1 Defines Variations:

   .1   Alteration or modification of design, quality or quantity of the Works, including:

      .1   addition, omission or substitution of any work

      .2   alteration of any kind or standard of any materials or goods to be used in the Works

      .3   removal from site of any work executed or materials or goods for the Works which *are* in accordance with the Contract (brought on site by the Contractor).

   .2   The imposition by the Employer of any obligations or restrictions in regard to the matters set out in Clauses 13.1.2.1 to 13.1.2.4 the addition to or alterations or omission of any such obligations or restrictions so imposed or imposed by the Employer in the Contract Bills in regard to:

      .1   access to the site or use of any specific parts of the site

      .2   limitations of working space

      .3   limitations of working hours

      .4   the execution or completion of the work in any specific order.

   Note: If sections are to be completed by specific dates in advance of the overall completion date, then the Sectional Completion Supplement must be used.

   .3   Specifically excludes omitting Contractor's measured work and the substitution of a Nominated Sub-Contractor to execute that work.

      Note: Must be a PC or provisional sum for a valid nomination.

      Any omissions must be genuine, otherwise the Contractor may claim loss of profit for the 'omissions'.

13.2 Subject to the Contractor's right of reasonable objection (Clause 4.1.1) Architect may issue AIs requiring Variations.

    Architect *may* sanction in writing any Variation made by the Contractor *not* pursuant to an AI.

    'No Variation required by the Architect or subsequently sanctioned by him shall vitiate this contract.'

This does *not* mean that the Architect may order any changes he likes and still maintain the Contract intact.

An *excessive* Variation ordered by an AI would not be regarded as a Variation under the Contract, thus payment would be on *quantum meruit.*

Lord Kenyon in *Pepper* v. *Burland* (1792):

'If a man contracts to do work by a certain plan, and that plan is so entirely abandoned that it is impossible to trace the Contract, and to what part of it the work shall be applied, in such case the workman shall be permitted to charge for the whole work done by measure and value, as if no Contract had ever been made.'

13.3   Architect to issue AIs regarding:

.1   expenditure of provisional sums included in the contract Bills,

.2   expenditure of provisional sums included in a NS/C (applies where nomination occurs from expenditure of a provisional sum as Clause 13.3.1).

Valuation of Variations

JCT 98 introduces three mechanisms by which a Variation can be valued. Two of these are relatively new and follow developments elsewhere (notably in the New Engineering and Construction Contract); these are the 13A Quotation and the Contractor's Price Statement (Alternative A). The third mechanism, of valuation by the Quantity Surveyor, is the traditional JCT approach (Alternative B).

13.4.1.2   Alternative A – Contractor's Price Statement.

13.4.1.2A1   The Contractor may submit a Price Statement in respect of a Variation (except, confusingly, where a quotation has been accepted under 13A, a '13A Quotation'). If this is the valuation route the Contractor must submit a statement to the Quantity Surveyor within 21 days of receipt of the Architect's Instruction. The statement should include the price for the work, based on the valuation rules of 13.5. It *may* also include the Contractor's direct loss and expense and any extension of time (under 13A quotation rules these are mandatory).

13.4.1.2A2   The Quantity Surveyor must notify the Contractor, within 21 days, of the statement's acceptability. If it is not accepted, reasons, in a similar format to the statement, must be given and an amended statement provided. The Contractor then has 14 days to accept the amendment; if not either party may refer to adjudication.

13.4.1.2A4.3 If not referred to adjudication then the variation is valued by the Quantity Surveyor under 13.4 Alternative B.

13.4   Alternative B
If no 13A Quotation is sought and no Alternative A Contractor's Price Statement submitted, or if the 13A Quotation, or Contractor's Price Statement or amended Price Statement is rejected, then the Variation must be valued by the Quantity Surveyor in accordance with the rules in 13.5.

13.4.1.2   .1   Valuation of Variations – by the QS – in accordance with Clause 13.5 (unless otherwise agreed by the Contractor and the Employer – this would constitute a Variation of the Contract and so the parties to the contract must agree to it). Also applies to work executed by the Contractor for which an Approximate Quantity is included in the BQ.

Note: The payment rules are applicable only to Variations as defined by the Contract. If in doubt, therefore, the Contractor should implement the authority for an AI procedure as Clause 4.2.

See *Myers* v. *Sarl* (1860): Drawings, even if prepared in the Architect's office and stamped, must be signed by the Architect to be valid authority to execute work. Possibly, however, the signature of a clerk preparing the drawings on behalf of the Architect would suffice.

(b)   Valuation of Variations and valuation of work for which an Approximate Quantity is included in the bills of quantities included in the Numbered Documents (as annexed to Agreement NSC/A) – Nominated S/C's work – to be in accordance with the relevant provision of NSC/C (Sub Contract form); unless otherwise agreed by NS/C and Contractor, with Employer's approval.

.2   Valuation of work constituting a PC Sum arising due to AIs regarding expenditure of a provisional sum for which the Contract has had a tender accepted (Clause 35.2) must be valued in accordance with that successful tender, not the normal Variation provision.

13.5 .1 Variations constituting additional or substituted work and work for which an Approximate Quantity is included in the BQ which can be valued by measurement – valuation to be of:

   .1 *similar* character, executed under *similar* conditions does not *significantly* change quantity of work in Contract Bills – rates and prices in BQ shall determine the valuation.

   .2 similar character, *not* similar conditions and/or significantly changes quantity of work in BQ – rates and prices in BQ to form basis for valuation (i.e. *pro rata* prices) with a fair allowance for the difference.

   .3 *not* of similar character to BQ – fair rates and prices.

Rather contentious – can varied work really be of similar character? What is a significant change in quantity? How much to *pro rata*? What is a fair valuation? Perhaps a guide is to read 'identical' for 'similar'.

   .4 Approximate Quantity is close to the quantity of work required – rates and prices in BQ to be used.

   .5 Approximate Quantity is not a reasonably accurate forecast of the quantity of work required – *pro rata* rates/prices to be used.

Clauses 13.5.1.4 and .5 apply only if the work is the same in all respects as that described in the BQ except for the quantity required. Hence, if the specification/conditions of the work change, the full valuation of Variations rules apply.

   .2 Variations requiring omission of work from BQ: valuation of work omitted to be at BQ rates.

   .3.1 Measurement to be in accordance with the principles used to prepare the BQ (SMM7).

   .2 Allowances to be included in respect of any appropriate percentages or lump sums in BQ (e.g. profit, attendances on NS/Cs).

   .3 Appropriate adjustment to be made for preliminaries. This provision does not apply in respect of compliance with an AI for the expenditure of a provisional sum for defined work (see SMM7 General Rules 10.1 to 10.6).

   .4 Where additions or substitutions cannot properly be valued by measurement, valuation shall comprise:

.1 Prime cost of the work plus percentage additions as set out by the Contractor in the BQ (usually separate percentages for labour, plant, materials). Prime cost – calculate in accordance with 'Definition of the Prime Cost of Daywork carried out under a Building Contract' current at the Date of Tender.

.2 Specialist trades definition of Prime Cost of Daywork to be used as appropriate.

(Electrical Contractors' Association – Electrical Contractors' Association of Scotland – Heating and Ventilating Contractors' Association.)

Vouchers stating:

time daily spent on the work,
workmen's names (usually including grade),
plant,
materials,

must be sent to Architect, or his authorised representative for verification not later than the end of the week following that during which the work is executed. Verification is usually by signature of the appropriate person – if in doubt send to the Architect.

.5 If a Variation *substantially* changes the conditions under which other work is executed, that other work may be valued in accordance with the rules for valuing Variations.

The provision also applies to other work subject to changes due to:

(a) compliance with an AI for expenditure of a provisional sum for undefined work

(b) compliance with an AI for expenditure of a provisional sum for defined work – where the work executed differs from the work described in the BQ

(c) execution of work for which an Approximate Quantity is included in the BQ – where the quantity executed is different from that stated in the BQ.

.6.1 Valuation of Performance Specified Work – to include allowance for addition/omission of any work involved in preparation and production of drawings, schedules or other documents.

.2 Valuation of additional/substituted work related to Performance Specified Work:

(a) work of similar character – BQ (or Analysis) rates or prices with allowances for changed conditions/quantities changed significantly from those in BQ or Contractor's Statement.

(b) no work of similar character in BQ or Contractor's Statement – a fair valuation.

.3 Valuation of omission of work relating to Performance Specified Work – at rates/prices in the BQ or the Analysis.

.4 Any valuations under Clause 13.5.6.2 and .3 must include any appropriate adjustments of preliminaries.

.5 Valuations of additional/substituted work relating to Performance Specified Work by daywork – the proviso to Clause 13.5.4 applies (vouchers, etc. – see above).

.6 If compliance with an AI requiring a Variation or expenditure against a provisional sum differs from the information in the BQ (both in regard to Performance Specified Work) and substantially changes the conditions under which any other work is executed, such work affected shall be treated as being the subject of an AI requiring a Variation and valued under the provisions for valuing Variations (Clause 13.5).

.7 Excluding additions, omissions and substitutions of work, the valuation of work or liabilities directly associated with a Variation which cannot be reasonably valued as above, shall be the subject of a fair valuation, e.g. part load delivery charges where the bulk of the items in question have already been delivered.

An allowance for disturbance to the regular progress of the works and/or for any direct loss and/or expense may be included in the valuation of Variations under Clause 13.5 only where it is *not* reimbursable under any other Contract Clause.

Usually, such 'disturbance cost' will be considered under Clause 26 – the provision of clause 13.5 being used only as a 'last resort'.

13.6 QS to allow Contractor to be present and take all necessary notes when QS is measuring work to value Variations.

13.7 Contract Sum to be adjusted to take account of all valuations of Variations.

Thus, where a Variation has been executed but not measured and valued, it is reasonable to include an 'on account' valuation in interim payments until formal valuation has been finalised.

Note: The valuation of Variations, howsoever arising, is to be executed by the QS (including provisional sum items) unless the
Employer and Contractor themselves agree something different.
The Architect is involved only where a Contractor claims for reimbursement of loss/expense due to the regular progress of the works
being materially affected (disturbed).

Clause 13A: Variation Instructions – Contractor's quotation in compliance with the instruction

By Clause 13.2.3 an Instruction requiring a variation may state that the treatment and valuation of the variation are to be in accordance with 13A. The Contractor has 7 days to disagree and then the Instruction must be valued by 13.4 Alternative A or B.

13A.1.1 The Instruction shall provide sufficient detail to allow the Contractor to provide a Quotation (a 13A Quotation).

13A.1.2 The Contractor has 21 days from receipt of the Instruction to submit his 13A Quotation (including 3.3A Quotations (NSC) in respect of any NSC) and that 13A Quotation remains open for acceptance for 7 days.

13A.2 The 13A Quotation is an all-inclusive price for the variation, including:

.1 the value of the work (including NSC work) plus allowances for preliminary items

.2 the adjustment to time for completion

.3 the amount in lieu of any ascertainment under Clause 26 of direct loss and expense

.4 the cost of preparing the 13A Quotation.

Where specifically requested the 13A Quotation shall provide indicative information statements on

.5 the additional resources required

.6 the method of carrying out the Variation.

13A.3.1 The Employer may accept a 13A Quotation and the Architect shall immediately confirm to the Contractor 'a confirmed acceptance'.

13.4 If the 13A Quotation is not accepted the Architect either instructs that the Variation is carried out, and valued via either Alternative A or B, or that the work is not to be carried out. If the work is not carried out the Contractor shall be paid a fair and reasonable amount in respect of the preparation of the Quotation.

Clause 14: Contract Sum

14.1 'The quality and quantity of the work included in the Contract Sum shall be deemed to be that which is set out in the contract Bills'.

Note: Clause 2.2 – The Bills of Quantities must be in accordance with SMM7. Any items not so included must be specifically noted with an indication of how measured, usually indicated thus:

'Notwithstanding SMM *Clause No.*, *work item* is measured *description of how measured for BQ.*'

Such a statement will usually occur in the Preambles section of the BQ.

Clause 14.1 is upheld in its expressed limitation of authority of the BQ by:

English Industrial Estates Corporation v. *George Wimpey* & *Co. Ltd* (1973)

Gleeson MJ Ltd v. *London Borough of Hillingdon* (1979).

14.2 Contract Sum fixed as a lump sum; the only adjustments permissible are those set out in the Conditions of Contract (e.g. valuation of Variations).

Subject to QS's preparation errors which are covered by Clause 2.2.2.2, any errors (arithmetic, etc.) are deemed to have been accepted by the parties and are non-adjustable.

Note:

(a) Provisions of Code of Procedure for Single Stage Selective Tendering, 1998.

(b) Professional responsibilities, especially negligence.

(c) JCT Standard Form is a Lump Sum Contract with provision for interim payments. Although the Lump Sum nature is diluted by items in the Bills of Quantities the Contractor agrees to carry out the work for the Contract Sum – not to carry out the work for the rates in the Bills. These rates are only for the valuation of variations.

Clause 15: Value added tax – supplemental provisions

15.1   Defines VAT – introduced by Finance Act 1972. Under the control of the Customs and Excise.

15.2   Contract Sum is always *exclusive* of VAT.

15.3   Where any items become *exempt* from VAT after the Base Date – Employer to pay the Contractor the loss of input tax he would otherwise have recovered.

Clause 16: Materials and goods unfixed or off-site

16.1   Unfixed materials/goods delivered to, placed on or adjacent to the Works and intended therefore may be removed only for use on the Works, unless removal has been agreed in writing by the Architect. Such consent not to be unreasonably withheld.

Unfixed materials/goods (as above) – value of which has been included in an Interim Certificate which has been paid are the property of the Employer. The Contractor is in the position of a bailee, expressly responsible for loss or damage to them; but subject to Clause 22B or C if applicable (fire, etc., Employer to insure).

Until an Interim Certificate, as denoted above, is honoured, the property in the goods remains with the appropriate party as per the sale agreement, governed by the Sale of Goods Act (usually the Contractor).

Following *Dawber-Williamson Roofing Ltd* v. *Humberside County Council* (1979); under sub-contract in which no express terms cover the passing of property in materials/goods on site, the title does not pass until the materials/goods are fixed.

See *Stansbie* v. *Troman* (1948): where items on site are owned by the Employer, common law requires the Contractor to take reasonable care to protect them from damage, theft, etc. This, probably, also applies to Sub-Contractors' items on site.

16.2   Applies to materials/goods stored off site – often applies to nominated items due to their high value. Once the Employer has paid the Contractor for them (clause 30.3) and so the property has passed to the Employer, the Contractor has responsibilities:

(a)   to remove them only for incorporation in the Works

(b)   for costs of storage, handling, insurance until delivered to the Works, whereupon Clause 16.1 applies

(c)   for loss or damage.

Note: S.16 Sale of Goods Act – property in unascertained goods cannot pass to the purchaser unless and until the goods are ascertained.

Naturally, many of these responsibilities are passed on by the Contractor to the Supplier or Sub-Contractor in most cases.

In the absence of any express terms in a building Contract, materials on site do not become the property of the Employer until they have been made part of the project; 'fully and finally incorporated into the Works' is a common phrase in this regard.

Clearly in the 1998 JCT standard form, this general principle is overridden by the express provisions of the Contract – Clause 16.

The requirement for listed items, either uniquely identified or not, of Clause 16 materials has implications for certificates and payments (Clause 30.3).

Clause 17: Practical Completion and defects liability

17.1 As soon as the Architect considers that Practical Completion of the Works has been achieved, he must issue a Certificate to that effect provided that the Contractor has provided any as-built drawings, etc. concerning Performance Specified Work in accordance with Clause 5.9. For all the purposes of the contract, Practical Completion is deemed to have occurred on the day named in that Certificate.

Practical Completion is not defined in the contract but is normally understood to be when the Works are complete for all practical purposes, any outstanding items of work being of only a minor or remedial nature such that they would not materially affect the proper functioning of the building.

Note: Practical Completion is distinct from and not applicable to the doctrine of Substantial Completion where, if a party can show that he has 'substantially performed' his obligations, he can successfully sue for the price under an entire or lump sum contract provided that he gives appropriate credit for any deficiencies in the whole, including uncompleted obligations.

See *Appleby* v. *Myers* (1867) – following *Cutter* v. *Powell* (1795) – Blackburn J:

'There is nothing to render either it illegal or absurd in the workman to agree to complete the whole, and to be paid when the whole is complete, and not till then.'

If a builder failed to complete an entire contract he could claim neither *quantum meruit* nor in equity.

Sumpter v. *Hedges* (1898) – A.L. Smith LJ:

'The law is that where there is a contract to do work for a lump sum, until the work is completed the price of it cannot be recovered.'

Dakin H. & Co. Ltd v. *Lee* (1916): Under a lump sum building contract, defects or omissions amounting only to negligent performance (not abandonment of the contract, etc.) did not preclude a successful claim for the contract sum less only the amount necessary to make the work accord with the specification.

46

Naturally the question is one of degree and for this rule to apply the omissions or defects should be of a relatively minor nature.

Hoenig v. *Issacs* (1952) – Denning LJ:

'It was a lump sum contract, but that does not mean that entire performance was a condition precedent to payment. Where a contract provides for a specific sum to be paid on completion of specified work, the courts lean against a construction of the contract which would deprive the contractor of any payment at all simply because there are some defects or omissions.'

Thus, it may be concluded that the JCT Standard Form 1998 Edition is a lump sum contract with provision for Interim Payments.

Note: following observations in *Dakin* v. *Lee* and *Hoenig* v. *Isaacs*:

(a) the builder cannot recover if he abandons the contract (subject to the express terms of the agreement)

(b) contracts which provide for retention money to be paid on completion might require entire performance in the strict sense – but, probably, in relation only to the retention releases rather than Interim Payments under the Contract.

(c) *J. Jarvis & Sons Ltd* v. *Westminster Corporation* (1970): Practical completion:

'What is meant is the completion of all the construction work that has to be done.'

(d) *H.W. Nevill (Sunblest) Ltd* v. *Wm Press & Son Ltd* (1980):

'I think that the word "practically" in Clause 15(1)[JCT 63] gave the Architect discretion to certify that William Press had fulfilled its obligation under Clause 21(1) where very minor, de minimis, work had not been carried out but that if there were any patent defects in what William Press had done, the Architect could not have given a Certificate of Practical Completion.'

17.2 The Appendix provides for the usual 6 months DLP to be varied.
Defects, shrinkages and other faults which appear within the DLP and are due to materials and workmanship not in accordance with the Contract must be:

(a) Specified on a schedule of defects by the Architect, to be delivered as an AI to the Contractor not later than 14 days from the expiration of DLP.

(b)   made good by the Contractor at his own cost (unless subject to an AI, with the Employer's consent, not to make good any such defects and to reduce the Contract Sum accordingly) and within a reasonable time.

17.3   Despite the provision of Clause 17.2, such defects etc., including damage caused by frost prior to Practical Completion, may be the subjects of an AI for their making good – Contractor to comply within a reasonable time and normally at his own cost. (Again, subject to a possible AI not to make good, and, then to reduce the Contract Sum accordingly.)

No such AIs may be issued after delivery of the schedule of defects or 14 days from the expiration of DLP.

This Clause enables individual defects, usually of a more major nature, to be required to be made good prior to the issue of the full defects list.

There is no provision for an interim defects list except by this AI provision but the issue of such a list is quite common in practice.

Thus, it is sensible and usual for the defects schedule to be prepared and issued as late as possible such that all defects arising may be properly included.

Any defects which appear after the expiration of DLP are not covered by the Contract and so any action for their making good would have to be at common law.

Note: Statutes of Limitations provisions.

17.4   When the defects specified by the AIs and/or the schedule of defects have, in the Architect's opinion, been made good, he must issue a Certificate of Completion of Making Good Defects.

This Certificate is a prerequisite for:

(a)   Final Certificate – Clause 30.8,

(b)   release of the balance of retention – Clause 30.4.1.3

17.5   Contractor must make good only frost damage which is due to frost prior to Practical Completion. If frost damage becomes apparent after Practical Completion the Architect must certify that such damage is due to frost which occurred prior to Practical Completion for the Contractor to be required to make it good under the Contract. If no such Certificate is issued, the Contractor is entitled to claim payment for the work involved.

Clause 18: Partial possession by Employer

18.1 Prior to the date(s) of issue of the Certificate(s) of Practical Completion, the Employer may wish to take possession of any part(s) of the Works. If the consent (not unreasonably withheld) of the Contractor has been obtained, then notwithstanding anything express or implied elsewhere in the Contract, the Employer may take possession. Upon such taking of possession, the Architect must issue a written statement to the Contractor, on the Employer's behalf, which identifies the part(s) of the Works taken into possession ('relevant part') and which specifies the date on which the Employer took possession ('relevant date').

In such instances, the following must occur:

.1 Practical Completion is deemed to have occurred and the DLP to have commenced on the relevant date for the relevant part, for the following purposes only:

(a) Clause 17.2 – schedule of defects

(b) Clause 17.3 – AIs re defects

(c) Clause 17.5 – damage by frost

(d) Clause 30.4.1.2 – Retention ($\frac{1}{2}$ release).

.2 Certificate of Completion of Making Good Defects of the relevant part to be issued at the appropriate time (see usual provisions regarding the issue of this Certificate).

.3 The obligation to insure the relevant part under Clause 22A (contract insuring) or Clause 22B.1 or 22C.2 (Employer insuring) terminates from the relevant date. Where Clause 22C applies, the obligation of the Employer to insure under Clause 22C.1 includes the relevant part from the relevant date.

.4 Liability to pay liquidated damages in respect of the relevant part ceases on the relevant date. Any such liability (Clause 24) in respect of the remainder of the Works which arises on or after the relevant date is reduced on a pro-rata basis. (The value of the relevant part pro-rata the Contract Sum.)

English Industrial Estates Corporation v. *George Wimpey & Co. Ltd* (1973) stresses the importance of the Architect's Certificate.

Clause 19: Assignment and sub-contracts

Note: the difference between assignment and sub-letting.

Assignment the transfer of one party's rights and obligations under a contract to a third party. The consent of the other party to the original contract is often required (e.g. debt factoring). Obligations cannot be transferred unless attached to some right(s).

Sub-letting where one party's contractual obligations are carried out on his behalf by a third party. The original parties to the contract retain full rights and responsibilities under that contract. Often, the third party will be in a contractual (Sub-Contract) relationship with the party for whom he carried out obligations vicariously.

19.1.1 Neither party may assign the Contract without the written consent of the other.

   .2 If the Appendix states that Clause 19.1.2 applies and if the Employer transfers his freehold or leasehold interest in the premises comprising the Works or grants a leasehold interest in such premises, at any time after Practical Completion, the Employer may assign to the new freeholder or leaseholder the right to bring arbitration or litigation proceedings (in the name of the Employer) to enforce any of the terms of this Contract.

    Any enforceable agreements regarding this Contract made between the Employer and Contractor prior to any such assignment remain intact and, via estoppel, are part of the assignment. (This is an expression of a basic legal premise regarding transfers of title 'nemo dat quod non habet'.)

Note:

(a) This provision for assignment by the Employer does not require the written consent of the Contractor.

(b) The rules of Limitation will remain.

(c) The assignment must be executed correctly – in accordance with the Contract, naming the assignee correctly, etc. (of particular importance to groups of companies).

19.2.1 Any Sub-Contractor which is not a NS/C is a Domestic Sub-Contractor.

.2 The Contractor must have the written consent of the Architect to sub-let any part of the Works. Such consent must not be unreasonably withheld.

Following Clause 2.1, the Contractor retains overall responsibility for the execution and completion of the Works in accordance with the Contract, irrespective of any sub-letting.

Note: LA common requirements to sub-let only to local firms – usually within a prescribed area.

There is no requirement for the Architect to approve domestic Sub-Contractors.

19.3   .1   Work measured in the Contract Bills and priced by the Contractor, where the BQ requires that work to be carried out by a person selected by the Contractor from a list contained in the BQ.

.2.1   The list must contain at least 3 persons. The Employer (or the Architect on his behalf) or Contractor may, with the consent of the other party, add persons to any such list prior to the execution of a binding Sub-Contract in respect of that work.

.2   If, prior to the execution of the Sub-Contract, less than 3 persons named in a list are able and willing to carry out the work:

either   the Employer and Contractor by agreement add further names to the list to make it at least 3 names long,

or   the work may be carried out by the Contractor who may sub-let it to a Domestic Sub-Contractor under Clause 19.2.

.3   A person selected from such a list is to be a Domestic Sub-Contractor.

Note: If a list is altered by having names added, such additions must be inserted in the BQ and initialled by the Employer and Contractor as they represent changes in the terms of the Contract.

19.4.1   Any domestic sub-contract must provide for the employment of that Domestic Sub-Contractor to be determined immediately that the employment of the main Contractor is determined under the contract.

.2   Any domestic sub-contract must provide that:

.1 Unfixed materials/goods on site of the Domestic Sub-Contractor must not be removed except for use on the Works, unless the Contractor has consented in writing to such other removal. The consent must not be withheld unreasonably.

.2 Where, in accordance with Clause 30, the value of unfixed materials/goods on site has been included in an Interim Certificate against which the Employer has *discharged* payment to the Contractor (usually, by making the payment), those materials/goods are deemed to be the property of the Employer. The Domestic Sub-Contractor may not deny that the property in the materials/goods has passed to the Employer.

.3 If the Contractor pays the Domestic Sub-Contractor for unfixed materials/goods on site prior to the Employer's properly discharging payment to the contractor for them, the property in those materials/goods, passes to the Contractor upon the payment's being made to the Domestic Sub-Contractor.

.4 The Operation of Clauses 19.4.2.1 to 3 is without prejudice to the provision of Clause 30.3.5.

Clause 19.4.2 is intended to overcome difficulties found to exist over the title to unfixed materials/goods on site in *Dawber-Williamson Roofing Ltd* v. *Humberside County Council* (1979), in instances where domestic sub-contracts include retention of title provision ('Romalpa Clauses' – *Aluminium Industrie Vaassen* v. *Romalpa* (1976)). Clauses 19.4.2.2 and unfixed materials/goods on site for which payment has been discharged under the provisions of the main Contract, even in cases where the Contractor does not have good title to those materials/goods to pass on – this is the objective of the 'not deny' stipulation in Clause 19.4.2.2.

Note: The requirement is to discharge payment not to make the payment for valid title to pass; this provision is to accommodate set-off, etc.

19.5.1 Provisions for Nominated Sub-Contractor – see Part II of the Contract.

The Contractor retains overall responsibility for executing and completing the Works in accordance with the Contract irrespective of work executed by NS/Cs unless the Contract states otherwise.

.2 Unless the Contractor is acting as a Nominated Sub-Contractor (Clause 35.2) he is not required to do any work which is to be carried out by a Nominated Sub-Contractor.

N.W. Regional Hospital Board v. *Bickerton* (1969): If a first nomination fails before the work under a PC Sum is completed, there is a duty upon the Architect to re-nominate.

Note: Non-compliance with the sub-letting provisions (Clause 19.2 etc.) is a ground for determination of the contractor's employment (Clause 27.1.4).

British Crane Hire Ltd v. *Ipswich Plant Hire Ltd* (1975): If there is a 'course of dealing' between the parties (frequent transactions probably using a standard contract) or if the terms of agreements are standard (e.g. national plant hire agreement), even if not specifically included in forming a contract, they may be implied to give the relationship 'business efficacy'.

Clause 20: Injury to persons and property and indemnity to Employer

See also Practice Note 22.

20.1 The Contractor to be liable for and must indemnify the Employer against any expense, liability, loss, claim or proceedings at statute or common law for personal injury or death due to the carrying out of the Works.

   The only exception is the extent to which negligence on the part of:

   (a) the Employer

   (b) any person for whom the Employer is responsible

   (c) any person(s) employed or engaged by the Employer to whom Clause 29 refers

   caused the injury/death. (Thus if the Employer's negligence is 30% to blame for the injury, the indemnity covers 70% of the Employer's loss.)

20.2 Similar to Clause 20.1, but:

   (a) refers to real or personal property

   (b) excludes the Works etc. as Clause 20.3

   (c) excludes damage to existing structures, contents etc. under Clause 22C.1 – Employer to insure

   (d) damage must be due to any negligence, breach of statutory duty, omission or default of the Contractor or persons for whom the Contractor is responsible. Includes all persons authorised to be on the site except:

      (i) the Employer

      (ii) persons employed, engaged or authorised by the Employer

      (iii) persons employed, engaged or authorised by any local authority

      (iv) persons as (iii) but employed, etc. by a statutory undertaker and who are executing work solely in pursuance of its statutory rights or obligations.

20.3.1 'Property real or personal' in Clause 20.2 subject to partial possession, excludes the Works, materials on site and work executed up to and including the earlier of:

(a) the date of issue of the Certificate of Practical Completion, or

(b) the date of determination of the employment of the Contractor (whether or not disputed) under Clauses 27 or 28 or 22C.4.3, if applicable.

Thus the Contractor is responsible for these items and so should effect appropriate insurance.

.2 If partial possession has occurred under Clause 18, the relevant part is excluded from the Works or work executed under Clause 20.3.1.

The purpose of indemnity is to protect against legal responsibility or to compensate; insurance proves a fund to enable the indemnifying party to make any payments which may arise. Insurance does not affect the obligations of the parties. Thus, it is usual for indemnity and insurance provisions to be considered together, the former apportioning risks and the latter dealing with the settlement of claims in respect of prescribed liabilities.

Law Reform (Married Women and Tortfeasors) Act, 1935:

Clause 20 covers tortious liability, thus the Act is of relevance. If there were joint (two or more) tortfeasors and only one were sued by a plaintiff and, thereby, had to pay damages, that tortfeasor could recover from the other, joint tortfeasors a contribution to the damages paid in proportion to the liabilities of each tortfeasor in respect of the tortious act.

Such does not apply if one joint tortfeasor is to indemnify the other(s); the party who is to indemnify the other(s) bears the full liability.

Here the contractor must indemnify the Employer.

See also:

Farr AE Ltd v. *Admiralty* (1953)
A.M.F. International v. *Magnet Bowling Ltd* & *G.P. Trentham Ltd* (1968).

Clause 21: Insurance against injury to persons or property

See also Practice Note 22

21.1.1.1   The Contractor must take out and maintain insurance in respect of liabilities placed upon him under Clauses 20.1 and 20.2, i.e.:

    (a) for personal injuries etc., arising from the Works (except if due to Employer's, etc. negligence)

    (b) for damage to real or personal property arising from the Works due to the negligence, etc. of the Contractor, etc. Note exclusions of persons causing the damage under Clause 20.2; item (d).

   .2   Insurance for injuries to or death of Contractor's employees which occur during the course of their employment must comply with the provisions of the Employers' Liability (Compulsory Insurance) Act 1975 including any amendments, etc., to that act.

      The minimum insurance cover required for all other claims under Clause 21.1.1.1 for any one occurrence or series of occurrences arising out of one event is stated in the Appendix.

Beyond the scope of the provisions of the Contract, the Contractor is required to comply with all statutory provisions, including those regarding insurance, to be effected by an employer.

21.1.2 The Contractor and Sub-Contractors must send documentary evidence to the Architect for inspection by the Employer that the insurances required by Clause 21.1.1.1 have been taken out and maintained whenever reasonably required to do so by the Employer.

      On any occasion, the Employer may require (not unreasonably or vexatiously) such documentary evidence to be relevant policy (policies) and premium receipt(s).

   .3 If the Contractor defaults in:

    (a) taking out the insurance, or

    (b) maintaining the insurance

as required under Clause 21.1.1.1., the Employer may effect the appropriate insurance and recover the premium amounts (paid or payable) from the Contractor:

(a) by deduction from payments to the Contractor under this Contract, or

(b) as a debt of the Contractor.

21.2.1 If the Appendix states that insurance under this Clause (21.2.1) may be required by the Employer, upon being so instructed by the Architect, the contractor must take out and maintain a Joint Names Policy (the joint names being those of the Employer and the Contractor) for the amount of indemnity stated in the Appendix. A policy of insurance taken out for the purposes of clause 21.2 should extend until the end of the DLP. The insurance is against:

'. . . any expense, liability, loss, claim or proceedings which the Employer may incur or sustain by reason of any damage to any property other than the Works and Site Materials caused by collapse, heave, vibration, weakening or removal of support or lowering of groundwater . . .'

due to the execution of the Works.
   Exceptions are injury or damage:

.1 for which the Contractor is liable under Clause 20.2

.2 due to errors or omissions in the design

.3 which reasonably can be foreseen to be inevitable (due to the nature or the manner of its execution (*Rylands* v. *Fletcher* (1868))

.4 for which the Employer should insure under Clause 22C.1, if applicable

.5 to the Works and Site Materials brought on to the site of the Contract for the purpose of its execution except insofar as any part or parts thereof are the subject of a Certificate of Practical Completion – the Employer then takes responsibility

.6 & .7 arising from war risks or the Excepted Risks (see definition of Excepted Risks in Clause 1.3)

.8 directly or indirectly caused by or arising out of pollution or contamination during the period of insurance. Then an exception to the exception – the Contractor is responsible for sudden, identifiable, unintended and unexpected incidents.

.9   which result in the Employer incurring costs or paying sums in respect of breach of contract, again with an exception to the exception, where such costs or payment would have occurred in the absence of any Contract

This insurance covers the Employer's potential liability as a joint tortfeasor – if the Contractor wishes to insure in respect of his potential liabilities in this regard, he must do so in addition to the insurance required by this Clause.

Under a joint names policy, the insurers cannot recover payment to one insured party from the other insured party where the latter's action caused the loss (i.e. subrogation cannot occur).

See *Gold* v. *Patman & Fotherington Ltd* (1958).

The insurance also will provide protection for the Employer where there is no negligence on the part of the Contractor, etc.

It would be prudent for the insurance under Clause 21.2.1 and that effected by the Contractor in respect of Clause 20.2 to be with the same insurer, thereby avoiding potential argument between insurers about which is liable.

.2 The Clause 21.1.1 insurance to be placed with insurers approved by the Employer. Contractor to send policy (policies) and premium receipt(s) to the Architect for deposit with the Employer.

.3 Amounts paid by the contractor to effect these insurances are to be added to the Contract Sum.

.4 If the Contractor defaults in respect of these insurance require- ments, the Employer may effect the insurance required.

21.3   Damage and injury caused by Excepted Risks are excluded from:

(a)   indemnity provisions of Clauses 20.1 and 20.2

(b)   insurance requirements of Clause 20.1.1

Thus, the Excepted Risks are assumed by the Employer.

Clause 22: Insurance of the Works

See also Practice Note 22.

If it is not possible to effect insurance in respect of all the 'Specified Perils' (see Clause 1.3 for definition), the matter should be resolved at Tender stage and the Contract amended accordingly (this should not be necessary).

Which of the three alternative Clauses will be used depends upon the circumstances:

22A	Contractor to insure (new building)
22B	Employer to insure (new building)
22C	Employer to insure – alterations/extensions to an existing structure

The explanation for the three alternatives is that the Employer may choose to insure new building work since an Employer may be better placed to secure an economic policy; and that only the Employer can insure existing buildings.

Each insurance comprises a Joint Names Policy for All Risks Insurance for the Works. Clause 22C insurance further comprises a Joint Names Policy for insurance of the existing structure and their contents owned by the Employer or which are the Employer's responsibility against loss or damage caused by the Specified Perils.

22.1 Which Clause, 22A, 22B or 22C is to apply must be stated in the Appendix.

Note: A new building, if attached to an existing building by a link bridge or similar connection, depending on the nature of the connection of the buildings, may be deemed to be an extension to the existing building for insurance purposes. Hence, care must be taken to ensure that the correct insurance is stipulated and effected otherwise, due to the nature of insurance contracts (*uberrimae fidae*), the 'apparent' insurance may be void.

22.2. Sets out relevant definitions of terms, namely:

All Risks Insurance – which gives cover against any physical loss or damage to work executed and Site Materials.

Costs of repairing, replacement or rectification are excluded in respect of:

.1 defects due to

.1 wear and tear

59

.2   obsolescence

.3   deterioration, rust or mildew;

.2   work executed or Site Materials lost or damaged through defective design or work execution, etc. of those items, and any work executed which is lost or damaged as a consequence of such failure where the lost or damaged work relied for support or stability on the items which failed;

Note: Site Materials do not feature in the second part of this sub-Clause.

See also: *D & F Estates Ltd* v. *Church Commissioners for England* (1989).

.3   loss or damage caused by or arising from

.1   any consequence of war, invasion etc., nationalisation, requisition, loss, destruction or damage to property by or under orders of any government, public, municipal or local authority;

.2   disappearance or shortage, only if revealed when an inventory is made or is not traceable to an identifiable event;

.3   an Excepted Risk.

If the Contract is in Northern Ireland two further exclusions apply:

.4   civil commotion;

.5   Terrorists acts – a terrorist is a member of or somebody acting for an organisation which is proscribed under the Northern Ireland (Emergency Provisions) Act, 1973; terrorism is violence for political ends including violence to scare the public or a section thereof.

Site Materials – all unfixed materials and goods delivered to, placed on or adjacent to the Works and intended for incorporation therein.

Note: No definition of temporary works is given but, normally, is defined as works executed by the Contractor to enable the permanent Works to proceed.

The insurance does not provide cover for temporary works, plant and similar items of the Contractor, etc.; insurance of such items should be effected by the Contractor (and Sub-Contractors).

All risks insurance policies are not standard. The Contract lists risks to be covered by the insurance, but it is probable that the working of policies will vary (as may the cover offered). It is essential to check that the requisite cover is afforded by the policy (perhaps particularly by scrutiny of exclusion clauses in the policy) prior to actually effecting the insurance. The JCT has checked to ensure that the required cover can be obtained.

22.3.1 The assured (Contractor under Clause 22A; Employer under Clauses 22B and 22C) must ensure that the Joint Names Policy(s) (under the applicable of Clauses 22A.1, 22A.3, 22B.1, 22C.1, 22C.2):

(a) provide for recognition of each NS/C as an insured under the relevant Joint Names Policy, or

(b) include a waiver by the insurers of their rights of subrogation (if any) against any NS/C.

Thus, under alternative (b), the insurer waives any right to sue any NS/C for damage caused by the NS/C and covered by (and thence reimbursed under) the Joint Names Policy.

The recognition or waiver for NS/Cs is in respect of loss or damage to the Works and Site Materials caused by Specified Perils under Clauses 22A, 22B and 22C.2. Where Clauses 22C.1 applies, the recognition or waiver applies to loss or damage to the existing structures and relevant contents caused by Specified Perils. Any such recognition or waiver applies up to and including the earlier of:

(i) date of issue of Certificate of Practical Completion of the Sub-Contract Works – NSC/C, Clause 2.11

(ii) date of determination of the employment of the Contractor (whether disputed or not) under Clause 27, 28, or 22.C.4.3 if applicable.

The provisions of Clause 22.3.1 also apply to any Joint Names Policy effected under the Contract where the party who was required to insure has not effected the insurance and the Joint Names Policy has, in consequence, been taken out by the other party (e.g. Employer has effected the insurance under Clause 22A.2).

22.3.2 The provisions of Clause 22.3.1 regarding recognition or waiver apply also to Domestic Sub-Contractors. Such recognition or waiver continues up to and including the earlier of:

(i) the date of issue of any certificate or document which states that the Domestic Sub-Contract Works are practically complete, or

(ii)   the date of determination of the Contractor's employment as Clause 22.3.1.

The provisions of Clause 22.3.1 do not apply where the Joint Names Policy is effected under Clauses 22C.1 or 22C.3.

Clause 22A: Erection of new buildings – All Risks Insurance of the Works by the Contractor

22A.1 Contractor to take out and maintain a Joint Names Policy for All Risks Insurance. The minimum scope of cover is denoted in Clause 22.2. The amount of cover required is the full reinstatement value of the Works plus any percentage to cover professional fees as stated in the Appendix.

Subject to Clause 18.1.3., the insurance must be maintained up to and including the earlier of:

(a) the date of issue of the Certificate of Practical Completion,

or

(b) the date of determination of the employment of the Contractor under Clause 27, 28 or 28A (whether the validity of such determination is contested or not).

Note:

(i) Especially during times of high inflation, it is essential to ensure that the cover is adequate for the full reinstatement value, as required – including Variations, etc.

(ii) Where an Extension of Time is awarded, it is essential to ensure that the insurance cover is extended up to and including the new Completion Date.

(iii) If the required All Risks Insurance cannot be obtained, prior to executing the Contract, the parties should either amend the definition of All Risks Insurance to accord with the cover which can be obtained or state the risks which are to be covered by the insurance in the Contract.

.2 Employer must approve the insurers with whom the contractor takes out the Joint Names Policy under Clause 22A.1. Contractor to send to the Architect, for deposit with the Employer:

(a) the Policy,

(b) the premium receipt for the Policy,

(c) any endorsements necessary to maintain the Policy, as required by Clause 22A.1, and

(d)   premium receipts for any such endorsements.

If the Contractor defaults in taking out and maintaining the insurance required under Clauses 22A.1 and 22A.2, the Employer may:

(a)   effect insurance by a Joint Names Policy against any risks regarding which the Contractor's default occurred,

(b)   recover the premium amounts (paid or payable by the Employer to effect the necessary insurance) from the Contractor:

   (i)   by deduction from payments to the Contractor under this Contract, or

   (ii) as a debt of the Contractor.

Note: Any amounts paid by the Employer to effect insurance under the Contract in the event of the Contractor's failure to effect the required insurance can be recovered from the Contractor either as a debt (so, ultimately, by litigation) or by set-off, etc., against monies due from the Employer to the Contractor under this contract; under JCT 80 such sums *cannot* be transferred from one project between the parties to another project. The usual means of recovering such sums is by contra-charge or set-off against monies due under Interim Certificates issued subsequently; set-off is a common law right.

22A.3.1   If the contractor maintains a Contractor's all risks insurance independently, such insurance will discharge the Contractor's obligations to insure under Clause 22A.1 provided that:

(a)   the scope of the cover is at least that required by Clause 22.2,

(b)   the amount of cover is adequate for the Work's full reinstatement value plus any percentage for professional fees (as Clause 22A.1), and

(c)   it is a Joint Names Policy in respect of the Works.

The annual renewal date of such Contractor's all risks policy must be supplied by the Contractor and stated in the Appendix.

The Contractor is not required to deposit such policy and premium receipts with the Employer provided that, when reasonably required to do so by the Employer, the Contractor can send documentary evidence that the policy is being maintained to the Architect for inspection by the Employer. Further, such inspection arrangements can be invoked by the Employer on any occasion in respect of the policy and premium receipts.

22A.3.2   The provisions of Clause 22A.2 (set-off, etc.) apply in respect of the Employer's rights to insure should the Contractor default in respect of taking out and maintaining the insurance required under Clause 22A.3.1 (contractor's annual all risks insurance).

22A.4.1   If any loss or damage which is covered by the insurance under Clauses 22A.1, 22A.2 or 22A.3 occurs, immediately upon discovering the loss or damage the Contractor must give written notice to the Employer and to the Architect stating the nature, extent and location of such loss or damage.

.2   The occurrence of such loss or damage must be ignored in calculating amounts payable to the Contractor under or due to the Contract.

.3   The Contractor must reinstate work damaged, replace or repair Site Materials which have been lost or damaged, remove and dispose of any debris and proceed with the proper execution and completion of the Works once any inspection required by the insurers regarding a claim made under the Joint Names policy under Clauses 22A.1, 22A.2 or 22A.3 has been completed. Again, the Contractor is required to work diligently to effect the repairs, etc.

.4   The Contractor must authorise the insurers to pay to the Employer all monies from the insurance under Clauses 22A.1, 22A.2 or 22A.3 regarding loss or damage. The authority is in respect of the Contractor and all those Domestic Sub-Contractors and NSCs who, under Clause 22.3, are recognised as insured under the Joint Names Policy. Those monies (less any element for professional fees) must be paid to the Contractor by the Employer as authorised by interim Certificate issued by the Architect (but not the provisions of Clause 22A.4.5).

.5   The only monies which the Contractor is entitled to receive in respect of the repairs etc. are those paid under the insurance. Thus, the Contractor must bear any costs incurred in excess of those monies paid under the insurance in the execution of the repairs, etc. (Any Variation, however, would be valued by the appropriate method and the Contract Sum adjusted accordingly.)

Clause 22B: Erection of new buildings – All Risks Insurance of the Works by the Employer

22B.1 The Employer is to effect the Joint Names Policy for All Risks Insurance as described in Clause 22A.1.

22B.2 (Not Local Authorities Form). As and when reasonably required to do so by the Contractor, the Employer must produce documentary evidence and receipts showing that the Joint Names Policy under Clause 22B.1 has been taken out and is being maintained. (The Contractor will probably desire to inspect the policy, any endorsements and the receipts for premiums paid.)

If the Employer defaults in effecting the requisite insurance, the Contractor may effect the necessary cover and any amounts paid or payable by the Contractor for such premiums must be added to the Contract sum.

As the Contractor is an insured under the Joint Names Policy effected by the Employer, it is not necessary for the Contractor to obtain separate insurance for the risks covered by the Joint Names policy. Any additional cover required by the Contractor (for hutting, plant, etc.) must be effected by the Contractor separately.

22B.3.1 As Clause 22A.4.1. (Refers to 22B.1 or .2 instead of 22A.1, .2 or .3.)

22B.3.2 As Clause 22A.4.2.

22B.3.3 As Clause 22A.4.3. (Refers to 22B.1 or .2 instead of 22A.1, .2 or .3.)

22B.3.4 As Clause 22B.4.4 (Refers to 22B.1 or .2 instead of 22A.1, .2 or .3 and to 22B.3.1 instead of 22A.4.1.)

22B.3.5 Restoration, replacement etc. work executed by the Contractor – paid for as a Variation required by an A.I. under Clause 12.2.

Note: (i) The Employer must ensure adequacy of cover during periods of inflation.

(ii) Although the mechanism for payment for repairs will be via Interim Certificates, as the repair work, etc. constitutes a variation, the monies available are not limited (as they are under Clause 22A) to the sum paid under the insurance.

Clause 22C: Insurance of existing structures – Insurance of Works in or extensions to existing structures

22C.1 The Employer must take and maintain a Joint Names Policy for:

(a) the existing structures,

(b) any relevant part under Clause 18.1.3 from the relevant date, and

(c) contents of such structures owned by or the responsibility of the Employer for their full cost of reinstatement, repair or replacement caused by one or more of the Specified Perils up to and including the earlier of:

    (i) the date of issue of the Certificate of Practical Completion, or

    (ii) determination of the employment of the Contractor under Clause 22C.4.3, 27, 28 or 28A (whether validity contested or not).

For the Contractor and all NS/Cs who in respect of Clause 22.3.1 are recognised as insured under the Joint Names Policy under Clauses 22C.1 or 22C.3, the Contractor must authorise the insurers to pay to the Employer all monies from that insurance in respect of any loss or damage.

22C.2 As Clause 22A.1 but includes possible determination of the employment of the Contractor under Clause 22C.4.3.

22C.3 (Not Local Authorities Form.)

As and when reasonably required to do so by the Contractor, the Employer must produce documentary evidence and receipts showing that the Joint Names Policy under Clause 22C.1 or 22C.2 has been taken out and is being maintained.

If the Employer defaults in effecting the requisite insurance under Clause 22C.1, the Contractor may effect the necessary cover for which purpose the Contractor has right of entry and inspection to make a survey and inventory of the existing structures and the relevant contents.

If the Employer defaults in effecting the requisite insurance under Clause 22C.2, the Contractor may effect the necessary cover. (No entry provisions as for default under clause 22C.1 are required as this insurance is in respect of the Works, which are in the possession of the Contractor.)

Amounts in respect of premiums paid by the Contractor under Clause 22C.3 must be added to the Contract Sum.

22C.4 If any loss or damage which is covered by the insurance under Clauses 22C.2 or 22C.3 occurs, immediately upon discovering the loss or damage the Contractor must give written notice to the Employer and to the Architect stating the extent, nature and location of such loss or damage, and:

    .1   The occurrence of such loss, etc. must be ignored in calculating any amounts payable to the Contractor under or due to the Contract,

    .2   The Contract must authorise the insurers to pay to the Employer all monies from the insurance under Clause 22C.2 or 22C.3, regarding loss, etc. as Clause 22C.4. Such authority is in respect of the Contractor, and all those Domestic Sub-Contractors and NSCs who, under Clause 22.3, are recognised as insured under the Joint Names Policy.

    .3   .1   Within 28 days of the occurrence of a Clause 22C.4 loss:

        (a) if just and equitable, the employment of the Contractor may be determined at the option of either party – notice by registered post or recorded delivery,

        (b) within 7 days from receipt of a determination notice, either party may give the other a written request to concur in the appointment of an Arbitrator (under Clause 41) to decide if such determination will be just and equitable.

        .2   In the event of determination of the Contractor's employment, Clauses 28A.4 and 28A.5 (except Clause 28A.5.5) are to be followed.

    .4   If determination of the Contractor's employment is inapplicable:

        .1   after any inspection required by the insurers is completed, the Contractor must make good and complete the Works, including replacement, etc. of damaged or lost Site Materials; and

        .2   such making good, debris disposal, etc. is regarded as a Variation required by an AI under Clause 13.2. (See notes to Clause 22B.3.5.)

Clause 22D: Insurance for Employer's loss of liquidated damages – Clause 25.4.3

This Clause provides an option for the Employer to insure against loss of liquidated damages where the Architect awards the Contractor an extension of time for loss or damage caused by the occurrence of one or more of the Specified Perils (flood, fire, etc.).

22D.1 The Appendix will state either:

(a)   Clause 22D insurance may be required, or

(b)   Clause 22D insurance is not required by the Employer.

If the Appendix states that Clause 22D insurance may be required, as soon as the Employer and the Contractor enter into the Contract, the Architect must inform the Contractor that either:

(i)   the insurance is not required, or

(ii)   the Contractor must obtain a quotation for the insurance.

The quotation must be for insurance:

(a)   on an agreed value basis (to avoid subsequent disputes over sum(s) payable – the insurers will satisfy themselves that the liquidated damages stated in the Contract are reasonable),

(b)   which provides for payment to the Employer of a sum calculated in accordance with Clause 22D.3.

The sum paid to the Employer is to compensate him for his loss of right to claim liquidated damages from the Contractor where the Architect has awarded an extension of time under Clause 25.3 due to loss/damage to the Works, etc. (including temporary buildings, plant, etc.) which was caused by the occurrence of one or more of the Specified Perils (flood, etc.), a relevant Event under Clause 25.4.3.

If the Contractor reasonably requires any additional information to obtain the quotation, the Architect must get it from the Employer.

If the Contractor has obtained a quotation:

(a)   the Contractor must send it to the Architect, as soon as practicable, and

(b)   the Architect must then promptly instruct the Contractor whether or not the Employer wishes to accept that quotation.

If the Contractor receives an AI that the quotation is to be accepted, the Contractor must:

(i)   forthwith take out and maintain the policy until the date of Practical Completion and

(ii) send the Policy, premium receipt plus any endorsements and their premium receipts to the Architect for deposit with the Employer.

.2 The sum insured is:

the rate of liquidated damages (as stated in the Appendix) for the period stated in the Appendix.

.3 Payment under the insurance is:

the rate stated in Clause 22D.2 (revised by application of Clause 18.1.4 – partial possession by the Employer) multiplied by the shorter period of either:

(a)   that stated in Appendix, or

(b)   that extension of time awarded by the Architect as referred to in Clause 22D.1 (Relevant Event is the Specified Peril(s)).

.4 Amounts spent by the Contractor in effecting Clause 22D insurance are added to the Contract Sum.

If the Contractor defaults in effecting the requisite Clause 22D insurance, the Employer may effect such insurance.

Clause 22FC: Joint Fire Code – compliance

The Joint Fire Code is designed to reduce the incidence of fire on construction sites.

   This clause is optional and only applies where it is stated in the Appendix.

22FC.2.1   The Employer shall comply with the Joint Fire Code, and ensure the compliance of his servants and agents. And via:

22FC.2.2   The Contractor is similarly bound.

22FC.3.1   If a breach of the Joint Fire code occurs and the insurer requires 'Remedial Measures', the Contractor shall carry these out.

22FC.3.2   If the Contractor fails to carry out any 'Remedial Measures' within 7 days the Employer may employ and pay others to do the work and withhold or deduct these from the Contractor or may be recoverable as a debt.

22FC.4   The Employer and the Contractor indemnify each other in respect of breaches of the Joint Fire Code.

22FC.5   If after the Base Date the Joint Fire Code is amended the cost of any compliance shall be added to the Contract Sum.

Clause 23: Date of Possession, completion and postponement

23.1.1 Possession of the site to be given to the Contractor on the Date of Possession (as Appendix). Upon being given possession of the site, the Contractor must begin work and proceed regularly and diligently with the execution of the Works and complete (Practical Completion) on or before the Completion Date (as Appendix – Clause 1.3).

Failure by the Employer to give the Contractor possession would be a breach of Contract: the Contractor would, thereby, be entitled to damages; time for completion of the Contract, if applicable, would be 'at large'. The Architect is not entitled to alter the Date of Possession.

.2 If the Appendix states that this Clause (23.1.2) is applicable, the Employer may defer giving possession of the site to the Contractor for a specified period of up to six weeks from the Date of Possession; if no period is specified, six weeks applies.

23.2 The Architect may issue AIs covering postponement of any work to be executed under the Contract.

Such AIs may:

(a) provide grounds for an Extension of Time (Clause 25.4.3.1)

(b) provide grounds for a direct loss/expense claim (Clause 26.2.5)

(c) provide grounds for the Contractor to determine his employment (Clause 28.1.3.4)

London Borough of Hounslow v. *Twickenham Garden Developments Ltd* (1970). Megarry J:

(a) 'The Contract necessarily requires the building owner to give the Contractor such possession, occupation or use as is necessary to enable him to perform the Contract.'

(b) Problem of interpretation of 'regularly and diligently' – use of programme to aid interpretation?

.3 .1 For purposes of the Work insurances, the Contractor possesses the site and the Works up to and including the date of issue of the Certificate of Practical Completion.

Subject to Clause 18 (partial possession by the Employer), the Employer is not entitled to take possession of any part of the Works until the date of issue of the Certificate of Practical Completion.

Thus, once the Employer has taken possession of the Works (or a relevant part), from the date of issue of the Certificate of Practical Completion, it is the Employer's responsibility to effect any insurance of the Works (or relevant part).

.2   Despite Clause 23.3.1, prior to the issue of the Certificate of Practical Completion, the Employer may use or occupy part or all of the site and the Works provided the Contractor has consented in writing to such occupation or use.

The Employer must notify the insurers under that applicable of Clauses 22A, 22B or 22C.2, .3 and .4 and obtain those insurer's confirmation that the intended use or occupation by the Employer will not prejudice (affect detrimentally) the insurance; this must occur before the Contractor consents to the Employer's use or occupation.

Once the insurer's confirmation has been obtained, the Contractor's consent may not be withheld unreasonably, i.e. the Contractor must be able to justify not giving consent to the desired occupation or use by the Employer.

.3   If either Clause 22A.2 or 22A.3 applies and the insurers have required payment of an extra premium in return for their confirmation under Clause 23.3.2, the Contractor must notify the Employer of the amount of that extra premium.

If the Employer still desires to occupy or use the site or Works:

(a) the extra premium must be added to the Contract Sum, and

(b) the Contractor must provide the employer with the receipt for that extra premium, if requested to do so by the Employer.

Clause 24: Damages for non-completion

24.1 If the contractor fails to complete the Works (Practical Completion) by the Completion Date, the Architect must so certify (this is, of course, subject to Extension of Time awards). If, after the issue of such a non-completion certificate, the Architect fixes a new completion date, that fixing cancels the non-completion certificate but the Architect may issue another non-completion certificate as necessary.

24.2 .1 Provided that the Architect has issued a Clause 24.1 certificate, and the Employer has informed the Contractor that he requires payment of liquidated and ascertained damages, the Employer may not later than 5 days before the final date for payment of the debt due under the Final Certificate either:

.1 require the Contractor to pay liquidated and ascertained damages or

.2 give notice (30.1.1.4, 30.8.3) that he will deduct from monies due the liquidated and ascertained damages.

Bramall & Ogden v. *Sheffield City Council* (1983): To be valid, liquidated damages provisions must be expressed so as to accord with the terms of the contract.

Jarvis John Ltd v. *Rockdale Housing Association Ltd* (1986): 'The "Contractor" can . . . be naturally and sensibly understood as referring to, in this case, J. Jarvis Ltd, its servants and agents, through whom alone it, as a corporation, can act.'

Note: Both Partial Possession by the Employer and/or Extensions of Time will mitigate the Contractor's liability.

The Employer may recover such sums from any payments to the contractor under this Contract only, not any other contract between the same parties.

.2 If an Extension of Time is awarded after a liquidated damages amount has been deducted, then there is provision for the Employer to repay such amount to the Contractor but without any interest thereon.

.3 Irrespective of whether the Architect issues any further certificate(s) of non-completion (under Clause 24.1), all written requirements of the Employer under Clause 24.2.1

(repayment/allowance by the Contractor of liquidated and ascertained damages LAD) remain effective unless withdrawn by the Employer.

Both the Contractor and the Employer must take care to ensure that the Employer's written 'notices' re LAD accord with the amounts due under non-completion certificates – this is of particular importance in computing the final account.

Note: Penalties for default are not recognised in English law. Thus, any sum stated as liquidated and ascertained damages must be a reasonable (genuine pre-estimate) estimate of the damage that the Employer would suffer if the project were to be completed late – see *Hadley* v. *Baxendale* (1854).

Dunlop Pneumatic Tyre Co. Ltd v. *New Garage & Motor Co. Ltd* (1915) – provides guidance as to what will constitute a penalty (and thereby be unenforceable but to which severance might be applied under Equity).

Peak Construction (Liverpool) Ltd v. *McKinney Foundation Ltd* (1971): If the Employer is in any way responsible for the Contractor's failure to meet the Completion Date, he cannot sue for liquidated damages.

He may still have a claim in respect of the Contractor's contribution to the late completion achieved.

Rayac Construction Co. Ltd v. *J.E. Lesser (Properties) Ltd* (1975) (High Court) – the failure of an Architect to issue a certificate re Clause 24.1 will not preclude a dispute regarding the validity of an Employer's counter-claim from going to Arbitration.

If any liquidated damages due to the Employer are not paid or otherwise allowed by the Contractor prior to the issue of the Final Certificate, the Employer's right to recover such amount from the Contractor is, it is submitted, terminated by the issue of the Financial Certificate. Such liquidated damages not recovered cannot, subsequently, be set off against any claims submitted by the Contractor.

This is an area of potential danger for the Architect – such a situation could lay the Architect open to a claim by the Employer for the liquidated damages he can no longer recover from the Contractor. Architects would be well advised to ensure the Employer has obtained all appropriate settlements, including liquidated damages, prior to issuing the Final Certificate.

The Employer maintains the right to recover liquidated damages by ensuring the claim is specified to the Contractor in writing as per Clause 24.2.1 and is agreed and acknowledged by the Contractor, preferably in writing, prior to the issue of the Final Certificate, such notice stating that the Employer may recover the appropriate sum at a later time.

Bell & Son (Paddington) Ltd v. *CBF Residential Care Housing Association* (1989): Under JCT 80 both a valid Architect's certificate of non-completion and written notice from the Employer are conditions precedent to the Employer's right to levy liquidated damages.

Rapid Building Group Ltd v. *Ealing Family Housing Association* (1984): A claim for unliquidated damages is limited to the losses which the Employer can prove have resulted from the late completion.

Tremloc Ltd v. *Errill Properties Ltd* (1987): Inserting 'nil' as the amount of liquidated damages in the Appendix removes all and any entitlement of the Employer to claim either liquidated or unliquidated damages.

Holme v. *Guppy* (1838): An employer loses the right to levy liquidated damages if he impedes the contractor in carrying out the work in such a way that an extension of time cannot be granted under the contract.

Clause 25: Extension of time

This Clause is closely associated with Clause 26, Clause 25 being time and Clause 26 being money.

A Clause 25 claim may arise on its own but a Clause 26 claim will be associated with a claim under Clause 25.

A successful Clause 25 claim obviates the liability for the Contractor to pay liquidated and ascertained damages for non-completion in respect of the time period for which the Extension of Time is granted.

Restrictions on the Contractor's right to fluctuations apply under Clauses 38.4.7, 39.5.7 or 40.7 if the Contractor is in default over completion.

25.1 Delay includes further delay for this Clause.

25.2 .1.1 Once it is reasonably apparent that the progress of the Works is being or is likely to be delayed, the Contractor must inform the Architect in writing stating the circumstances causing the delay and noting the Relevant Event(s) (see Clause 25.4) that he considers applicable.

i.e. The Contractor must indicate the cause of the delay as denoted by Clause 25.4 for any claim to be considered. If the claim is not covered by any of the reasons given by Clause 25.4, it will *not* be valid for consideration of an Extension of Time.

.2 If the Contractor's notice under Clause 25.2.1.1 refers in any way to an NS/C, the Contractor must give that NS/C a copy of the notice.

.2 The Contractor must, about each Relevant Event specified by him in a notice, either within the notice or as soon as possible in writing specify:

.1 the expected effects

.2 the estimated extent of expected delay in completing (if any) – also to any NS/Cs named in Clause 25.2.1.2.

.3 The Contractor to give further written notices (including to NS/Cs as named) as necessary or requested by the Architect to keep up to date particulars of delays (Clause 25.2.2.1, and time involved estimates (Clause 25.2.2.2) including any material changes to them.

25.3.1 On receipt of information as required regarding alleged delays if, in the Architect's opinion:

.1   any of the events stated by the Contractor are Relevant

      and

.2   the completion of the Works is likely to be delayed beyond
      the Completion Date due to these Relevant Events,

then the Architect shall give the Contractor a written Extension of
Time by fixing such later Completion Date as he considers to be
fair and reasonable.

In fixing such later Completion Date, the Architect must state
the points covered by the following two sub-clauses:

.3   which Relevant Events he has taken into account

      and

.4   the extent to which, if any, he has taken into account omis-
      sion Variations issued subsequent to the last fixing of the
      Completion Date.

The Architect shall, if practicable and if he has sufficient information, fix the new Completion Date within 12 weeks, i.e. not later
than 12 weeks from receipt of the Contractor's notice.

If the period from the Architect's receipt of the Contractor's
notice and the existing Completion Date is less than 12 weeks, he
shall fix the revised Completion Date before the existing Completion is reached.

If the Architect decides that no revision to the Completion Date
is warranted, he must so inform the Contractor within the 12 week
time period.

Note: It is the Architect's opinion regarding the Relevant Event's
      applicability and time involved which governs the award of
      any Extension of Time. The Contractor's written notice is a
      prerequisite for an extension as is the provision of his esti-
      mate of the delay involved. The Architect does not have to
      apportion any Extension of Time awarded between the
      Relevant Events causing the delay, except as required by
      Clause 26.3 to assist in the calculation of loss/expense
      awards under Clause 26.1.

(1963 Edition – no period to be given by Contractor. Architects
usually delayed awards until Practical Completion and then made
awards retrospectively (except for strikes). The new system seems
preferable.)

Amalgamated Building Contractors Ltd v. *Waltham Holy Cross UDC*
(1952): no longer applicable but applied to 1963 Edition.

Miller v. *LCC* (1934):

(a) If the Architect fails to grant an Extension of Time when he should have done so, liquidated damages are *not* recoverable.

(b) It appears that if the cause(s) of delay is within the control of Employer or Architect and no applicable Extension has been granted by the Completion Date last determined, the Completion Date is 'at large', i.e. the Contractor must complete within a reasonable time.

(c) If the Contractor disagrees with any Extension award (e.g. no Extension) the Contractor may invoke Arbitration prior to Practical Completion (or alleged Practical Completion) – Clause 41.

25.3.2 The first award by the Architect must be an extension of time, i.e. fix a Completion Date later than that originally specified in the Contract.

Subsequent consideration and awards may lead to the Completion Date being revised to one earlier than an extended date due to omission Variations being taken into account in the date revision, provided that the relevant omissions were due to AIs issued after the last revision of the Completion Date.

Despite omissions, the *Architect* cannot revise the Completion Date to one earlier than that specified in the original Contract – expressed by Clause 25.3.6.

25.3.3 Within 12 weeks from the date of Practical Completion the Architect must, in writing to the Contractor, do one of the following:

.1 revise the existing Completion Date to a later one if fair and reasonable so to do in the light of Relevant Events, whether or not those Relevant Events have been notified to the Architect by the Contractor. (Note actual wording – permits review by the Architect.)

.2 revise the existing Completion Date to an earlier one – if reasonable due to omissions subsequent to last fixing of the Completion Date.

.3 confirm the existing Completion Date.

This provision permits the Architect to make an Extension of Time award in the absence of the Contractor's written notice, which therefore in this instance is *not* a pre-requisite to an Extension,

.4 provided that

.1   the Contractor uses his best endeavours to prevent
      delays. See: *Greater London Council* v. *Cleveland Bridge &*
      Engineering Co. (1986).

This last point often (1963 Edition) has been interpreted
to mean that the Contractor must try and make up the time
of delays. This is *not* so and accounts for the new 12 weeks
award rule being seen as of major importance.

Note: 'Best endeavours' must be interpreted in the context
          of the contract and the circumstances under which
          the contract was made.

IBM UK v. *Rockware Glass Co.* (1980): 'Ask what steps a
prudent, determined and reasonable (person), acting in his
own interests and desiring to achieve that result, would take'.

Transfield Pty v. *Arlo International* (1980): (Australian)
(Consider all the circumstances to determine what can be
expected of a reasonable person using his 'best endeavours'.)
The standard required is, 'what is reasonable in the circumstances, having regard to the nature, capacity, qualifications
and responsibilities', of the person with the obligation?

.2   the Contractor must do all that is reasonably required to
      the satisfaction of the Architect to proceed with the
      Works.

.5   The Architect to notify every NS/C of each revision to the
      Completion Date.

.6   The Architect may not revise the Completion Date to one
      earlier than that stated in the Appendix.

Balfour Beatty Building Ltd v. *Chestermount Properties Ltd*
(1993):

(a) The underlying purpose of the completion date, exten-
    sion of time and liquidated damages provisions is to
    determine the aggregate period of time within which the
    Works ought to be completed.

(b) Clause 25.3.1 allows the Architect to extend time for
    Relevant Events which occur before the completion date.

(c) Under Clause 25.3.3, the Architect may adjust the com-
    pletion date retrospectively at any time up to 12 weeks
    from the date of Practical Completion.

(d) If the Architect instructs Variations whilst the Contractor
    is in culpable delay (Contractor's fault), the Architect

should award an extension of time for the Variation on a 'net' basis, i.e. from the existing completion date.

25.4   List of Relevant Events:

.1   *Force majeure* – Act of God; man-made events beyond the control of the parties.

.2   *Exceptionally* adverse weather conditions – normal adverse weather (predictable) is assumed to be incorporated into the programme – note effects of location, time of year and stage of job – presumably covers excessive heat and drought as well as cold, rain, snow and frost. The term 'inclement weather' was removed after exceptionally good weather was held to be adverse but not inclement and contractors could not obtain extension of time in periods of drought.

Note: site weather records – diary and use of Meteorological Office data to establish validity of claims. It is often good practice to record delays due to weather and actual weather conditions on a daily basis and to agree such records at the monthly site meeting.

Walter Lawrence and Son Ltd v. *Commercial Union Properties (UK) Ltd* (1984): held that the effects of exceptionally adverse weather (under JCT 63, so exceptionally inclement weather was considered) should be assessed regarding the time at which the works affected are executed, rather than at the time indicated on the programme (if any) for their execution.

.3   Loss or damage from 'Specified Perils' (flood, etc.).

Note: despite the requirements of Clause 22, separate notice is also required under this Clause.

.4   Civil commotion, strike or lock-out – at the site, Sub-Contractor's or Supplier's premises, or transport directly connected with the work execution.

.5   Compliance with AIs:

.1   discrepancies, divergencies (Clause 2.3)
discrepancies, divergencies – Performance Specified Work (Clause 2.4.1)
Variations (Clause 13.2) except confirmed acceptance of a 13A Quotation
provisional sums (Clause 13.3), except for AIs re:
     Performance Specified Work
     postponement (Clause 23.2)
     antiquities (Clause 34)

NS/Cs (Clause 35)
NSups (Clause 36)

.2   inspection and testing where items do comply with the Contract (Clause 8.3).

.6   .1   where an Information Release Schedule has been provided, failure of the Architect to comply with Clause 5.4.1 (release of information)

.2   failure of the Architect to comply with Clause 5.4.2. (release of further information).

The provision of the programme is of significance, especially if together with a schedule of key dates for the release of information (*not* a Contract Document), in a practical sense.

Contractually, the specific written application for the required information is essential, and so even a master programme denoting key dates is insufficient, 'provided that such application was made on a date which having regard to the Completion Date was neither unreasonably distant from nor unreasonably close to the date on which it was necessary for him to receive the same'.

Following *Glenlion Construction* v. *The Guinness Trust* (1987), where the contractor is required to provide a programme, there is no implied term that the employer (and representatives, etc.) should perform the contract so as to enable the contractor to execute the work as per that programme. The contract was JCT 63 and the programme showed work to achieve completion ahead of the contract completion date. Under JCT 98, the contractor must complete the Works by or on the Completion Date, and so the ruling in Glenlion remains appropriate.

This situation is concerned with Extension of Time. The obligation upon the Architect to provide the necessary information at the requisite time is now greater due to the requirement of the programme provision by the Contractor but the Contractor must still properly apply, in writing, for the necessary information.

.7   Delay by NS/Cs or NSups which the Contractor has taken all practicable steps to avoid or reduce.

Note: Contractor is obliged to minimise the delays as far as possible, not to make up time.

City of Westminster v. *J. Jarvis & Sons Ltd and Peter Lind Ltd* (1969): the question concerning the Completion Date when latent defects were discovered in piles – S/C was in breach; not delay as Practical Completion of the piling S/C had been achieved.

Trollope & Colls Ltd v. *N.W. Metropolitan Hospital Board* (1973): Completion Date on a phased project. If actual dates given these hold good despite delays to previous phases of the project.

Failure of the Employer to give the Contractor possession of the site is not a ground for an Extension of Time. In such circumstances the time for completion will be 'at large' with the consequent effect upon the possibility of deducting liquidated damages. (Any attempt to require the Contractor to complete by the original Completion Date would necessitate the Contractor working faster than he intended and would, thereby, change the terms of the offer (Tender) and its acceptance and is, therefore, invalid.)

.8.1 Delay or failure to execute work not forming part of the Contract by the Employer himself or by persons employed or otherwise engaged by the Employer as referred to in Clause 29 or the failure to execute such work. Historically this was known as the 'Artists and Tradesmen' Clause. These are direct to the Employer and, thus, outside the scope of the Contract, as such, and the Contractor's control.

It is probable that the Employer could recover damages from the 'Artists and Tradesmen' if they are independent contractors.

.2 Delay in or failure to supply goods and materials which the Employer is contractually obliged to supply.

.9 By *the Government of the United Kingdom*, exercising any statutory power after the Base Date and, thereby, adversely affecting the Contractor's procuring necessary labour, goods, fuel and energy essential to the proper execution of the Works.

.10.1 The Contractor's inability to secure such labour as is *essential* to the proper carrying out of the Works.
This must be:

(a) beyond the contractor's control, and

(b) not reasonably foreseeable at the Base Date.

.2 Reproduction of Clause 25.4.10.1 but for goods and materials.

In such instances the Contractor must exercise reasonable foresight – this will be applicable to his ordering and, thereby, pass on any liability to S/Cs and Suppliers. He must endeavour to procure the requisite specified items as far as possible.

The items must be essential to the Works – alternatives must be considered as must alternative methods of working and types of construction.

.11 Execution delay or failure by a Local Authority or Statutory Undertaker in pursuance of a statutory obligation.

This does not cover work done outside the scope of a statutory obligation. In such a case the Statutory Undertaker would be acting as an ordinary Nominated or Domestic S/C. If acting as an NS/C the delay, etc., is covered Clause 25.4.5.1, if acting as a Domestic S/C the delay must be borne by the Contractor.

.12 Failure by the Employer to give, in due time, access to the site via property which he possesses and controls in accordance with the Contract Bills and/or Drawings, provided any required notice for such access has been given by the Contractor to the Architect, or the Architect and Contractor have themselves agreed an access provision (via the Employer's property). Access includes both ingress to and egress from the site – perhaps particularly relevant to conversion projects.

Following *Porter* v. *Tottenham UDC* (1915), the Contractor assumed the risks of access obstructions caused by third parties.

Due to the wording of this Clause, the ruling would appear still to be applicable –

'in the possession and control of the Employer', is the vital issue.

Following *Hounslow Borough Council* v. *Twickenham Garden Developments Ltd* (1971), in considering an extension of time, the Architect is entitled to take into account any amount by which the Contractor is ahead of programme and to reduce any extension accordingly.

.13 Employer's deferment of giving possession of the site to the Contractor where the Appendix states that Clause 23.1.2 applies.

.14 Execution of work covered by an Approximate Quantity in the Contract Bills where such an Approximate Quantity is not a reasonable accurate forecast of the work actually required.

What constitutes reasonable accuracy is open to interpretation hence, a practical approach would be to agree the acceptable variability of Approximate Quantities at the outset (preferably specifying such variability in the tender documents).

.15 Delay due to a change in Statutory Requirements occurring after the Base Date which necessitates alteration/modification to any Performance Specified Work.

.16 Threatened or actual acts of terrorism or actions of authorities in dealing therewith.

.17 Compliance or non-compliance by the Employer with Clause 6A.1 (Statutory obligations, notices, fees and charges).

.18 Delay arising from a suspension by the Contractor of the performance of his obligations under the Contract to the Employer pursuant to Clause 30.1.4 (Rights of suspension of obligations by Contractor).

Clause 26: Loss and expense caused by matters materially affecting regular progress of the Works

This Clause is, generally, viewed as the money element of the Clauses 25 and 26 combination. As such, the items covered whereby the disturbance of the regular progress of the Works (or any part thereof) is material and causes direct loss and/or expense to the Contractor, are less in number and scope than the Relevant Events specified in Clause 25.

The matters listed by Clause 26 as being a basis for a claim are limited to those Relevant Events considered to be within the control of the Employer and his agents (e.g. the Architect) and deferment of the Employer's giving possession of the site to the Contractor if Clause 23.1.2 applies.

26.1 The Contractor's making written application to the Architect is a condition precedent for a claim under this Clause. The written application must state that he has incurred or believes he will incur direct loss and/or expense in the execution of the Contract.

Note: It must not be possible under the Contract for the Contractor to be reimbursed under any other provision as the loss is (or will be) attributable solely to the disturbance of the regular progress of the Works, that the progress is *materially* affected and that this has been caused by one or more of the matters listed in this Clause (clause 26.2).

.1 The Contractor's application must be made as soon as the delay is apparent, is likely or should have reasonably become apparent. It may affect the entire Works or any part thereof.

F.G. Minter Ltd v. Welsh Health Technical Services Organisation (1980) (Court of Appeal):

'In the building and construction industry "cash flow" is vital to the Contractor and delay in paying him for the work he does naturally results in the ordinary course of things in his being short of working capital, having to borrow capital to pay wages and hire charges and locking up in plant, labour and materials capital which would have been invested elsewhere.'

The case was concerned with the 1963 Edition of the JCT Contract (with certain amendments), the major point at issue being the inclusion of finance charges in a claim for direct loss and/or

86

expense (Clause 24 of the 1963 Edition), which, it was decided, would be applied to the period from the Contractor incurring the loss/expense to the giving of the loss/expense notice. Under JCT 80, a Contractor may give notice about future anticipated loss/expense, under Clause 26.1, and so, it is submitted, interest may apply for the incurrence of the loss/expense to the point of settling the claim; probably certification of the appropriate sum, that being the juncture at which a debt is created. Interest may continue to accrue after Practical Completion.

The result of the appeal hearing contains two points of relevance to any claims under Clause 26 of the 1980 Edition which would, by implication, apply:

(a) Direct loss and/or expense may include finance charges or interest, and

(b) It is (probably) open to the Contractor to make a single application for the reimbursement of the capital sum together with the interest thereon from the date the expenditure was incurred until the date of certification. (This is because the Contractor's written application under Clause 26.1 must state that he has incurred *or is likely to incur* direct loss and/or expense in the execution of the contract . . .) However, he is probably safer to make a series of claims in respect of a continuing loss/expense, e.g. finance charges.

In *Rees & Kirby Ltd* v. *Swansea City Council* (1985), the Court of Appeal took the view that finance charges should be calculated on the same basis as a bank overdraft or deposit interest in that interest should be compounded at quarterly intervals.

Such finance charges must be distinguished from interest on a quantified debt which is paid late, as this interest is payable only where contract terms so permit, otherwise it is **not** payable under common law.

Following *Croudace Ltd* v. *London Borough of Lambeth* (1985), failure of the Architect to include finance charges when settling contractor's claims is a breach of contract rendering the Employer liable.

Note: The Contractor is, obviously, regarded as a construction expert (professional) in the context of foreseeing delays and the causes thereof.

.2 The Contractor must supply the Architect with any information in support of the application to allow the Architect to decide its validity and effects. This is qualified by:

(a) reasonableness

(b) the request for information by the Architect.

.3 The Contract must submit details of loss and/or expense to the Architect or Quantity Surveyor to enable that party to ascertain the extent of the loss. This, again, is qualified by:

(a) reasonableness

(b) the request of the Architect/QS for the details.

As soon as the Architect is of the opinion that the Contractor's application has some validity, he (or the QS under his instruction) must from time to time ascertain the amount of such loss and/or expense incurred by the Contractor.

It should be noted at this juncture that this Clause does *not* cover the more general cases where circumstances have changed thereby increasing the cost of executing the Works.

26.2 The matters giving rise to a claim are specified:

.1 where an Information Release Schedule has been provided, failure of the Architect to comply with Clause 5.4.1 (Information Release Schedule)

.2 failure of the architect to comply with Clause 5.4.2 (Provision of further drawings or details)

See *Trollope & Colls Ltd* v. *Singer* (1913).

.2 Opening up and testing where the items are found to be in accordance with the Contract.

.3 Discrepancies between Contract Drawings and Bills.

.4.1 The execution of work not forming part of this Contract by the Employer himself or by persons employed or otherwise engaged by the Employer as referred to in Clause 29, or the failure to execute such work (Artists and Tradesmen).

.2 Supply failure by the Employer.

.5 AIs issued under Clause 23.2 re postponement.

.6 Failure of the Employer to give access.

.7 AIs for Variations and expenditure against provisional sums, except for a Variation for which the Architect has given a confirmed acceptance of a 13A Quotation.

.8 Execution of work against an inaccurate Approximate Quantity.

.9 Compliance or non-compliance by the Employer with obligation to ensure the Planning Supervisor (and/or Principal Contractor) carries out duties under CDM Regulations.

.10 Suspension by the Contractor pursuant to his rights to suspend (30.1.4).

26.3 The Architect is to state in writing to the Contractor his Extension of Time award regarding certain specified Relevant Events to the extent that this is necessary to determine the amount of any direct loss and/or expense.
(re Clauses 2.3; 13.2; 13.3; 23.2; 25.4.5.2; 25.4.6; 25.4.8; 25.4.12.)

26.4 .1 Provides for the Contractor to pass on to the Architect any written application, properly executed under Clause 4.38.1 of NSC/C, of a Nominated S/C for direct loss/expense due to delays.

The Architect, if satisfied about the delays and causes (or the QS on his behalf), shall ascertain the amount of the loss/expense.

.2 If necessary to do so to determine the amount of such loss, the Architect must inform, in writing, both the Contract and NS/C of any relevant Extension of Time in respect of certain specified Relevant Events.
(S/Contract Clauses 2.6.5.1 (in reference to main Contract 2.3; 13.2; 13.3; 23.2) and 2.6.5.2; 2.6.6; 2.6.8; 2.6.12; 2.6.15.)

26.5 Any awards to be added to the contract Sum as awarded – see also Clause 3.

26.6 The provisions of Clause 26 are without prejudice (i.e. have no effect upon) any other rights and remedies the Contractor may possess. For example, claim for breach by the Architect's non-supply of requisite information – see *Trollope & Colls Ltd* v. *Singer* (1913).

In *International Minerals and Chemical Corporation* v. *Karl O. Helm AG and Another* (1986) '. . . the surviving principle of legal policy is that it is a legal presumption that in the ordinary course of things a person does not suffer any loss by reason of the late payment of money. This is an artificial presumption but is justified by the fact that the usual loss is an interest loss and that compensation for this has been provided for and limited by statute. It follows that a plaintiff, where he is seeking to recover damages for the late payment of money, must prove not only that he suffered the alleged additional special loss and that it was caused by the defendant's default but also that the defendant had knowledge of the facts or circumstances which make such a loss a not unlikely consequence of such a default. In the eyes of the law, those facts or circumstances are deemed to be special . . .'

Special damages occur under the second branch of the rule in *Hadley* v. *Baxendale* (1854) as being within the knowledge (reasonable expectation) of the defendant party at the time of contracting.

As cash flow is acknowledged as being vital in the construction industry, any delay in making a payment, quite obviously, would involve the Contractor's incurring additional finance charges (interest), such additional costs should be recoverable.

Clause 27: Determination by Employer

27.1 Notices to be in writing and delivered by:

(a) actual delivery,

(b) special delivery, or

(c) recorded delivery.

If notices are sent by special delivery or recorded delivery, they are deemed to be received 48 hours after the date of posting, excluding Saturdays, Sundays and public holidays, and subject to proof to the contrary of delivery.

.2.1 If, prior to Practical Completion, the Contractor defaults in any (or all) of:

.1 *wholly* suspending execution of the Works without reasonable cause i.e. the Contractor may suspend execution of part of the Works but not the entire project

2 failure to proceed regularly and diligently with the Works

West Faulkner Associates v. *London Borough of Newham* (1994): '. . . regularly and diligently should be construed together and that, in essence, they mean simply contractors must go about their work in such a way as to achieve their contractual obligations. This requires them to plan their work, to lead and to manage their workforce, to provide sufficient and proper materials and to employ competent tradesmen so that the works are fully carried out to an acceptable standard and that all time, sequence and other provisions of the contract are fulfilled.'

.3 refusal or persistent failure to remove defective items which *materially* affect the Works and regarding which the Architect has given him written notice to remove

.4 failure to comply with Clause 19.1.1 or 19.2.2

(Assignment and Sub-contracts)

.5 failure to comply with the CDM Regulations.

The Architect may give written notice specifying the default to the Contractor.

Note: The delivery method of the written notice is specified. Sub-Clauses .1, .2, .3 represent repudiatory breach of

Contract thus, for any of these, the Employer has a common law right to '*determine the contract*', as discussed by the House of Lords in *Photo Productions Ltd v. Securicor Transport Ltd* (1980)

.2   If the Contractor continues the default for 14 days from receipt of the notice the Employer *may* on or within 10 days of the expiry of the 14 days serve notice of determination of the Contractor's employment upon the Contractor. Determination of the Contractor's employment takes effect on the date of receipt of the further notice (the notice of determination of the Contractor's employment).

Contractor repeats default
specified on a separate
future occasion – 14 days
period *not* applicable

14 days	10 days

Architect aware
of Contractor's
default prepares
notice specifying

Contractor continues
with default specified

Contractor receives
Architect's notice
re: default

Employer determines
Contractor's
employment under
contract

.3   If

(a) the Contractor stops the specified default(s), or

(b) the Employer does give the further notice (determination of the Contractor's employment) under Clause 27.2.2 and the Contractor repeats a specified default, the Employer may, within a reasonable time of such repetition of the specified default, issue a further notice and, thereby, determine the employment of the Contractor (effective as the date of the further notice).

.4   Notices under Clauses 27.2.2 and 27.2.3 must not be given unreasonably or vexatiously. (In such circumstances, the notices would be invalid.)

See also: *Hill* v. *London Borough of Camden* (1980)

Note: The determination of the Contractor's employment applies to that Contract on which the default occurs and the action is taken only.

.3.1 If the Contractor:

(a) makes a composition or arrangement with his creditors or becomes bankrupt, or (if the Contractor is a company)

(b) makes a proposal for a voluntary arrangement for a composition of debts or scheme of arrangement (i.e. seeks voluntary liquidation due to insolvency), or

(c) has a provisional liquidator appointed, or

(d) has a winding up order made, or

(e) passes a resolution for voluntary winding up (except for amalgamation or reconstruction of the company), or

(f) an administrator or administrative receiver is appointed under the Insolvency Act 1986 (or re-enactments)

i.e. if the Contractor is wound up voluntarily or compulsorily (other than for reconstruction or amalgamation of a company), then:

.2 the Contractor must inform the Employer in writing – under (a) or (b),

.3 the Contractor's employment is determined automatically – under (c) to (f) the employment may be reinstated if agreed by the Employer and Contractor,

.4 where Clause 27.3.3 does not apply and in the absence of any novation agreement under Clause 27.5.2.1, the Employer may determine the Contractor's employment at any time, in accordance with the provisions of the Contract, by notice to the Contractor – the determination being effective on the date of receipt of such notice (see Clause 27.1).

.4 Any corruption over securing contracts with the Employer by the Contractor entitles the Employer to determine the Contractor's employment. The extensive scope that this clause attempts to give – 'this or any other contract' – may not be workable, depending on the provisions in those other contracts.

.5 where Clause 27.3.4 applies:

.1   subject to Clause 27.5.3, the Employer is not bound to make any further payments and the Contractor is not bound to execute any further work from the date when the Employer first could give notice of determination of the Contractor's employment.

.2   Clause 27.5.1 applies until either of the following occurs:

.1   the Employer makes a '27.5.2.1 agreement' with the Contractor on the continuation, novation or conditional novation of this Contract – the Contract is then subject to the terms set out in the 27.5.2.1 agreement.

.2   the Employer determines the Contractor's employment – Clause 27.6 or 27.7 applies.

.3   Prior to a 27.5.2.1 agreement or determination of the Contractor's employment under Clause 27.3.4, the Employer and Contractor may make an interim arrangement for work to be executed. Subject to Clause 27.5.4, any rights of set-off by the Employer do not apply to payments due under that interim agreement.

.4   From the date of the Employer's initial entitlement to determine the Contractor's employment, he may take reasonable measures to safeguard Site Materials, the site and the Works. The Contractor must allow such measures to be executed and the Employer may set off any associated costs from monies due or which become due to the Contractor, including any monies under a 27.5.2.1 agreement.

27.6   If the employment of the Contractor is determined under Clause 27.2.2; 27.2.3; 27.3.3; 27.3.4; 27.4 and not reinstated:

.1   The Employer may employ another Contractor to complete the Works who may use all the items (huts, plant and materials) on or adjacent to the Works and intended for use thereon. This provision includes the use by the new Contractor of the original Contractor's *hired* plant. The consent of the owner of plant, etc. to such use must be obtained by the Employer.

.2.1   Except in the case of insolvency of the Contractor as under Clause 27.3.1, if the Architect or Employer requires, within 14 days of the date of determination, the contractor must assign all Sub-Contracts and supply Contracts in respect of the Work to the Employer (without any payment for such assignment). The Suppliers and Sub-Contractors who have Contracts so assigned may object to any further assignment (e.g. to the new Contractor) and would, probably, then remain in direct relationship to the Employer in respect of the new Contractor as Artists and Tradesmen and items supplied by the Employer.

.2 Except in the case of insolvency of the contractor as Clause 27.3.1, the Employer may pay any Supplier or S/C for items (work done or goods, etc., *delivered*) for the Works for which the Contractor has not discharged payment. This is in addition to similar provisions in respect of NS/Cs. Any such payments may be contra-charged against the Contractor.

.3 The Architect must inform the contractor in writing (usually, but not necessarily, and AI) when he requires the contractor to remove all his temporary buildings and plant (including hired items).

If the Contractor does not remove the items within a reasonable time from the Architect's written request, the Employer may:

(a) remove the items in question (he is not liable for any loss or damage thereto under the Contract)

(b) sell the items

(c) hold the proceeds of the sale to the credit of the Contractor after deduction of all costs incurred in the removal and sale.

.4.1 Subject to Clauses 27.5.3 and 27.6.4.2, the provisions of the Contract regarding ('normal') further payments release(s) of Retention to the Contractor shall not apply; i.e. the particular provisions concerning actions following determination of the Contractor's employment are those which do apply. However, the contractual rights of the Contractor to receive monies due are preserved expressly; all monies due up to 28 days prior to the initial date on which the Employer could give notice of determination (Clause 27.3.4 applies) or the date of determination ('automatic' – Clause 27.3.4 does not apply).

.2 Upon completion of the Works and making good of defects (under Clause 27.6.1, subject to 'waiver' as Clause 17), an account in accordance with Clause 27.6.5 must be prepared and set out in a statement by the Employer or in a certificate issued by the Architect within a reasonable time.

.5.1 Amount of expenses properly incurred by the Employer, including those under Clause 27.6.1, plus any direct loss/damage caused to the Employer by the determination;

  .2 Amount of any payment discharged in favour of the Contractor;

  .3 Total amount which would have been payable for the Works in accordance with the Contract.

.6 If there is a difference between the amount determined under Clauses 27.6.5.1 plus 27.6.5.2 and the total under Clause 27.6.5.3, the difference is a debt, payable as appropriate.

.7.1 Within 6 months of determination of the Contractor's employment, the Employer may elect not to complete the Works; if so, the Employer must so notify the Contractor in writing.

    Within a reasonable time from such notification, the Employer must send a statement of account to the Contractor, setting out:

  .1 Total value of work properly executed at the date of determination and any other amounts properly due to the Contractor, all as per the Contract;

  .2 Amounts of any expenses properly incurred by the Employer plus any direct loss/damage caused to the Employer by the determination.

    After accounting for amounts previously discharged to the Contractor, any balance is a debt payable by/to the Contractor.

  .2 If, after 6 months from the determination of the Contractor's employment, the Employer has neither begun operation of Clause 27.6.1 nor given notice under Clause 27.7.1, the Contractor may serve written notice on the Employer which requires the Employer to state whether Clauses 27.6.1 to 27.6.6 are to apply and, if not, for the Employer to prepare a statement under Clause 27.7.1 for submission to the Contractor.

.8 The Employer's rights under Clauses 27.2 to 27.7 are, expressly, without prejudice to (have no effect upon) the Employer's other possible rights and remedies.

Illustration

Contract Sum	£800 000
Value of Work executed	£250 000
Certificates (honoured)	£220 000

Thus the value of work to be completed by the original Contractor, A, on the basis of the contract Sum (Assuming no Variations or Fluctuations)	= £550 000

Final account of Contractor B in completing the original Contract Work	= £600 000

Excess price of B over A to complete	£50 000
Delay costs (A off to B on site)	say £10 000
	£60 000

Deduct:		
Sale of A's plant	£25 000	
less sale expenses	£5 000	£20 000
		£40 000

Deduct:	
Owed to Contractor A but not certified prior to determination	£30 000

Debt owed by Contractor A to Employer	£10 000

i.e. the Employer is Contractor A's creditor for £10 000.

Determination of Contractor's Employment

Check-list for action by the Architect

1   .1   Information is received that the Contractor is in financial difficulty/has become insolvent/is in liquidation.

.1   Establish the validity of the information taking care not to spread what may be a false rumour.

.2   Obtain the precise situation from the Receiver or Liquidator, if appointed.

.3   Establish the situation on the site and take the necessary steps to prevent unauthorised removal of items – see Clause 27.4.2.

The actions outlined above should not be undertaken too forcefully until the full facts have been established, i.e. keep a 'low profile'.

.2   If not determined automatically due to liquidation, etc., the Contractor's employment may be determined under Clause

27.2 due to non- or inadequate performance. Such determination is at the option of the Employer.

2   The Employer should be fully advised of the situation and his instructions obtained.

   .1   Action should be discussed and recommended regarding:

      .1   Liaison with the Receiver.

      .2   Safety and security measures (including any already implemented).

      .3   Re-insurances.

      .4   Completion of the Works – by the same or another Contractor?

      .5   Assignment of Sub-Contracts and supply Contracts.

      .6   Direct payments to Sub-Contractors and Suppliers.

   .2   It should be agreed who is to take any action considered necessary.

3   The other consultants should be informed of the situation as it develops and the agreed action. Particularly, the Quantity Surveyor will be involved.

   .1   The items covered in 2 above should be considered.

   .2   An accurate record of site conditions, progress, stocks, etc., is vital; photographs and an audit are usually required.

   .3   The procedure for completing the project should be agreed and implemented.

4   Written requirements (usually by AIs) should be issued to the contractor. Typically these will include such contractual matters as:

Non-removal of plant, materials and hutting on site which will be required for the completion of the project,

Assignment of Sub-Contracts and Supply Contracts,

and such 'practical matters' as:

Closure of the site,
Safety and security measures,
Safety of site documents and records.

5   It may be necessary for another Contractor to be employed to implement any necessary safety and security measures.

6   The Clerk of Works will have much useful information – this should
     be fully documented.
        If on site, he must be instructed regarding who may have access
     to the project and what items, if any, are to be delivered, removed
     and any work to be carried out.

7   If the Contractor is in liquidation, the Liquidator should be con-
     sulted to determine his intentions regarding the Contractor's
     future. It may well be advantageous (if possible) to arrange for the
     Contractor's employment to be reinstated (Clause 27.5).

8   It may be necessary for the Quantity Surveyor to prepare tender
     documents for the completion of the project.

9   A full discussion should take place between the consultants and the
     Employer should be advised of the most appropriate policy to
     obtain a suitable Contractor to complete the project:

     .1   Competitive Tender?

     .2   Negotiation?

10   Arrangements should be made to enable the new (or reinstated)
     Contractor to recommence site working as soon as possible.

11   Assignments of Sub-contracts and Supply Contracts must be for-
     malised, if required, from the original Contractor to the Employer.
     Unless the Sub-Contractors or Suppliers object (proviso of reason-
     ableness) the assignment is most usefully effected from the origi-
     nal to the new Contractor. (Clause 27.6). The Notice from the
     Architect requiring assignment must be given within 14 days of the
     date of determination.

12   When the plant of the original Contractor is no longer required,
     arrangements must be made for its removal from the site – this
     may involve sale of the plant by the Employer.

13   Once the project has been completed and the accounts verified
     (reasonable time from Completion) the final account between the
     Employer and the original Contractor must be settled. Note the
     provisions regarding set-off by the Employer in respect of the addi-
     tional costs incurred by him to achieve completion.

Reference: Ginnings, A.T. – 'Determination of Employment under the
Standard Forms of Contract for Construction Works' – *The Quantity Surveyor*, February 1978, pp 97–101.

Clause 28: Determination by Contractor

28.1 All notices under Clauses 28.2.1, .2, .3, .4 and 28.3 must be in writing and delivered by

(a) actual delivery, or

(b) special delivery, or

(c) recorded delivery.

For (b) and (c), notices are deemed to be delivered 48 hours after the date of posting (excluding Saturdays, Sundays and public holidays), subject to evidence to the contrary.

.2.1 If the Employer defaults in one or more of:

.1 not paying the amount properly due to the Contractor in respect of any Certificate and/or any VAT due as per the VAT Agreement

.2 interfering with or obstruction of the issuing of any Certificate due under the Contract. This will include payments, Practical Completion Certificates.

R.B. Burden Ltd v. *Swansea Corporation* (1957): The Clause does not include instances where certificates have been issued negligently and, thus, incorrectly. The interference or obstruction must be 'active'. A list of such instances would be very lengthy but would include:

(a) The Employer directing the Architect as to the amount to be certified.

(b) The Employer refusing access to the site for the Architect for the purpose of issuing a certificate.

.3 assigning the Contract without the written consent of the Contractor (Clause 19.1.1)

.4 fails to comply with CDM requirements

the Contractor *may* give the Employer a notice specifying the default(s) ('specified default or defaults').

.2 The execution of the whole or the majority of the uncompleted Works is suspended for a continuous period of the period stated in the Appendix) prior to the date of Practical Completion.

The reasons for suspension of the Works are:

.1.1 where an Information Release Schedule has been provided, failure of the Architect to comply with Clause 5.4.1

.2 failure of the Architect to comply with Clause 5.4.2.

.2 AIs regarding

(a) Clause 2.3 – discrepancies or divergencies

(b) Clause 13.2 – Variations

(c) Clause 23.2 – postponement

unless due to the Contractor's (including servants', agents', etc.) negligence/default except NS/Cs, Employer and any persons employed or engaged by the Employer. (See *Jarvis John Ltd* v. *Rockdale Housing Association Ltd* (1986).)

.3 Delay in the execution of work not forming part of this Contract by the Employer himself or by persons employed or otherwise engaged by the Employer as referred to in Clause 29 or by the Employer in supplying goods direct.

.4 Failure of the Employer to give access to the site (see Clause 25.4.12).

The Contractor *may* give the Employer a notice specifying the event(s) ('the specified suspension event or events').

.3 If the Employer continues a specified default or a specified suspension event is continued for 14 days from receipt of the notice (Clause 28.2.1 or 28.2.2), the Contractor may, on or within 10 days of the expiry of the 14 days, serve (a further) notice of determination of the Contractor's employment (which takes effect on the date of receipt of that further notice).

.4 If

(a) the Employer stops the specified default(s), or

(b) the specified suspension event(s) ceases, or

(c) the contractor does not give the further notice (of determination of employment under Clause 28.2.3)

and

(d) the Employer repeats a specified default, or

(e) a specified suspension event is repeated whereby the progress of the Works is/is likely to be materially affected, the Contractor may, within a reasonable time of such repetition, issue a further notice and, thereby, determine the Contractor's employment (effective on the date of the further notice).

.5   A notice of determination must not be given unreasonably or vexatiously; otherwise it would be invalid.

J.M. Hill Ltd v. *London Borough of Camden* (1982): for determination to be unreasonable, it must be totally unfair and, almost, smacking of sharp-practice. The fact that one reason for giving the determination notice was the Contractor's wish to protect himself against a potentially crippling loss and that it would result in his getting profit does not make the action of the Contractor unreasonable.

The situation, stated in *J.M. Hill* was expanded and reinforced in *Jarvis John Ltd* v. *Rockdale Housing Association* (1986): it was acknowledged that a contractor had a right to protect his interests in deciding what actions to take – '. . . it is only where there is a gross disparity between the benefit to him and the burden to the Employer that the exercise of his right even approaches unreasonableness.'

Thus, it appears that the burden of proof to establish that a contractor has acted 'unreasonably or vexatiously' is a heavy one.

.3.1 (Essentially, a replication of Clause 27.3.1) but in respect of insolvency, etc. of the Employer.)

If the Employer:

(a) makes a composition or arrangement with his creditors or becomes bankrupt, or (if the Employer is a company), or

(b) makes a proposal for a voluntary arrangement for a composition of debts or a scheme of arrangement (i.e. seeks voluntary liquidation due to insolvency), or

(c) has a provisional liquidator appointed, or

(d) has a winding up order made, or

(e) passes a resolution for voluntary winding up (except for amalgamation or reconstruction of the company), or

(f) an administrator or administrative receiver is appointed under the Insolvency Act 1986 (or re-enactments)

i.e. If the Employer is wound up voluntarily or compulsorily (other than for reconstruction or amalgamation of a company)
then:

.2 the Employer must inform the Contractor in writing – under (a) or (b).

.3 (in any situation covered by Clause 28.3.1) the Contractor may determine his employment under the contract by notice to the Employer; the determination takes effect on the date of receipt of the notice.

From the occurrence of any event under Clause 28.3.1 to the date at which a determination notice takes effect, the obligation of the Contractor to carry out the work as per Clause 2.1 is suspended. (This provision is useful to protect the Contractor from further possible losses by having to carry out work for an Employer known to be insolvent; the provision acts to preserve the 'status quo' from the point of the Employer's 'liquidation'.)

.4 If determination of the Contractor's employment under Clauses 28.2.3, 28.2.4 or 28.3.3 occurs (specified defaults, suspension events, 'liquidation', etc.), provided the employment has not been reinstated, Clauses 28.4.1 to .3 apply. The accrued rights and remedies of the parties, including those under Clause 20 (indemnity) remain intact up to the point of the parties having removed their huts, etc. from the site (the removal process is covered by the continuation of rights and liabilities).

Only the provisions of Clauses 28.4.2 and 28.4.3 apply regarding payments and releases of Retention.

.1 The contractor shall, as soon as possible, and with due regard to the safety of persons and property (following Clause 20), remove his temporary buildings, plant, materials (including Site Materials), etc. and ensure that Sub-Contractors do likewise. (Any items, notably Site Materials, in which the property has passed to the Employer, must remain on site.)

.2 The Employer must pay to the Contractor all Retention deducted by the Employer prior to the determination of the Contractor's employment within 28 days of the determination. Such payment is subject to any rights (set-off) which accrued to the Employer prior to the determination.

.3 As soon as practical, the Contractor must prepare an account in accordance with Clauses 28.4.3.1 to 28.4.3.5, including amounts in respect of NS/Cs:

.1 total value of work properly executed at the date of determination (calculated as per the Contract provisions as if the employment had not been determined) plus all other amounts due to the Contractor under the Contract

.2 (a) direct loss/expense due to delays (Clause 26)

(b) direct loss/expense due to antiquities (Clause 34.3)

.3 reasonable cost of removal of Contractor's and Sub-Contractors' items (Clause 28.4.1)

.4 direct loss/damage caused to the Contractor by the determination

.5 cost of materials or goods (including Site Materials) correctly for the Works for which the Contractor has paid or must pay. On payment therefor, the property in such materials/goods passes to the Employer.

Any balance, after allowing for sums previously paid/discharged in favour of the Contractor, must be paid by the Employer to the Contractor:

(a) within 28 days of the Contractor's submitting the account to the Employer, and

(b) without deduction of the any Retention.

The account, and its settlement, applies to this Contract only (it must not be an amalgam of several contracts between the same Employer and Contractor).

.5 The Contractor's rights under Clauses 28.2 to 28.4 are, expressly, without prejudice to (have no effect upon) the Contractor's other possible rights and remedies.

It may be possible for the Contractor (and Sub-Contractors and Suppliers) to claim loss of profit in respect of a project subject to determination by the Contractor, the loss of profit being on the uncompleted contract work and forming part of the direct loss/expense claim (Clause 28.3.4).

Wraight Ltd v. *P.H.T. Holdings Ltd* (1968)

Lintest Builders Ltd v. *Roberts* (1980): the Contractor is not in breach of his obligation to proceed with the works 'regularly and diligently' hereby because some defective work has been done; however, this situation, probably, is a matter of degree; necessary remedial work (re remedy defects, etc.) may be set-off by the Employer against any sums due to the contractor as under Clause 28.2 there are accrued rights of the Employer to have defective work rectified – such rights arise when the defective work is done.

Clause 28A: Determination by Employer or Contractor

28A.1 .1   Before the date of Practical Completion, if the carrying out of the whole, or the majority of the Works is suspended for the relevant continuous period stated in the Appendix, the Employer or the Contractor may determine the Contractor's employment under the Contract.

The determination is effected by written notice actually delivered (or by registered post or recorded delivery) that unless the suspension of work is terminated within 7 days of the receipt of the notice, the Contractor's employment is determined 7 days after the receipt of the notice.

The reasons for the suspension of work are:

.1 force majeure

.2 loss/damage to the Works caused by any one or more of the Specified Perils

.3 civil commotion

.4 Als under Clauses 2.3, 13.2 or 23.2 due to negligence/default by a statutory undertaker, etc.

.5 hostilities involving the UK, whether war declared or not

.6 terrorist activity.

.2   The entitlement of the Contractor to give notice (Clause 28A.1.1) of determination of employment due to the occurrence of Specified Peril(s) (Clause 28A.1.1.2) does not apply if the Contractor, etc. caused the loss/damage through negligence/default. (The entitlement does apply if the loss/damage is caused by negligence/default of

(a) the Employer,

(b) Employer's employee(s), 'Artists and Tradesmen',

(c) local authority/statutory undertaker executing work under statutory obligation(s).)

.3   A notice of determination (under Clause 28A.1.1) must not be given unreasonably or vexatiously.

.2 Upon determination of the Contractor's employment (Clause 28A.1.1), only the provisions of Clauses 28A.3 to 28A.6 apply regarding payments and releases of Retention.

.3 The Contractor shall, as soon as possible, and with regard to the safety of persons and property (following Clause 20), remove his temporary buildings, plant, materials (including Site Materials), etc. and must ensure that his sub-contractors do likewise. (Any items, notably Site Materials, in which the property has passed to the Employer, must remain on site – see Clause 28A.5.4.)

.4 Payment of Retention by the Employer to the Contractor:

   (a)   One half of the Retention deducted by the Employer prior to the determination – within 28 days of the date of the determination;

   (b)   Remainder – as part of the account under Clause 28A.5 and subject to any rights of deduction (set-off) which accrued prior to the determination.

.5 Within 2 months (calendar) of the date of the determination, the Contractor must provide all documents (including those relevant to NS/Cs and NSups) to the Employer which are necessary for the preparation of the account of settlement under this Clause.

Subject to the Contractor's providing such documents, the Employer must prepare the account promptly, comprising:

   .1   total value of work properly executed at the date of determination (calculated as per the Contract provisions as if the employment had not been determined) plus all other amounts due to the Contractor under the Contract

   .2   (a) direct loss/expense due to delays (Clause 26),

         (b) direct loss/expense due to antiquities (Clause 34.3)

   .3   reasonable cost of removal of Contractor's and sub-contractors' items (Clause 28A.3)

   .4   cost of materials or goods (including Site Materials) correctly for the Works for which the Contractor has paid or must pay. On payment therefor, the property in such materials/goods passes to the Employer

   .5   direct loss/damage caused to the Contractor by the determination.

         Any balance, after allowing for sums previously paid/ discharged in favour of the Contractor, must be paid by the Employer to the Contractor:

(a) within 28 days of the Employer's submitting the account to the Contractor, and

(b) without deduction of any Retention.

.6   If determination has occurred due to (Clause 28A.1.1.2) loss/damage occasioned by a Specified Peril(s) caused by negligence/default of the Employer, etc., that loss/damage caused to the Contractor must be included in the account of settlement under Clause 28A.5.

.7   The employer must inform the Contractor, and the relevant NS/Cs, in writing, of sums included in the account of settlement (under Clause 28A.5) attributable to each NS/C.

Clause 29: Works by Employer or persons employed or engaged by Employer

Often, this Clause is termed 'Artists and Tradesmen'.

29.1 Provided that the Contracts Bills contain sufficient information for the Contractor to execute the Works in the presence of the Employer or others directly engaged by him executing work which is not part of the Contract, the Contractor must permit such Artists and Tradesmen to execute their work on the site.

29.2 If the Contract Bills do not provide such requisite information regarding work of Artists, and Tradesmen, the Employer may, but with the contractor's consent (not unreasonably withheld), arrange for the execution of the work concerned.

Any extra costs thereby caused to the contractor are recoverable – see Clause 13.1.2 – restrictions imposed by Employer; Clauses 25.4.8.1 and 26.4 – delays and direct loss/expense.

29.3 Every Artist and Tradesman is the responsibility of the Employer, expressly in respect of injury to persons and property and Employer's indemnity (Clause 20).

Clause 30: Certificates and payments

There are now two ways in which the payment due to the Contractor on an interim payment can be arrived at and certified for payment. The traditional way is where the Quantity Surveyor values the work or by acceptance of the figure stated in the Contractor's Application (Clause 30.1.2).

30.1 .1.1 The Architect must issue Interim Certificates stating the amount due to the Contractor from the Employer (see Clause 30.1.3 and Appendix).

Note: The monthly intervals are *calendar months*, not lunar months, thus 12 payments per year (Section 61 of the Law of Property Act, 1925).

The Contractor is entitled to payment of the sum stated due in an Interim Certificate within 14 days from the date of its issue (period of honouring).

If the Employer fails to pay, in addition to the amount not paid he shall pay simple interest at 5% over the Base Rate of the Bank of England. This interest payment does not affect the Contractor's right in respect of suspension or determination.

.1.2 The Employer has a right to deduct retention.

.1.3 Within 5 days of the issue of an Interim Certificate the Employer shall give a written notice to the Contractor detailing the payment to be made.

.1.4 At least 5 days before the date for payment the Employer may give a written notice detailing any amounts to be withheld.

.1.5 If the Employer does not give the two written notices the Employer can make no deductions.

.1.6 An advance payment may be made to the Contractor if the Appendix so provides, and if a bond is required only after the bond has been provided.

.2 The Employer may set-off monies due from the Contractor to him under the Contract against any amounts due to the Contractor under an Interim Certificate. This right is qualified by the provisions regarding Retention in respect of direct payments to NS/Cs (set-off against the retention held against the Contractor, as Clause 35.13.5.3.2).

110

30.1 .1.3 Interim Certificates must be issued on the dates provided for in the Appendix.

.2.1 Interim Valuations:

(a) to be made by the Quantity Surveyor

(b) whenever the Architect considers them to be necessary to determine the amount to be stated due in an Interim Certificate.

Note: If fluctuations are to be recovered under the NEDO Formulae provisions, Clause 40.2 requires this Clause to be amended to require an Interim Valuation to be made as a pre-requisite for the issue of each Interim Certificate.

Following the ruling in *Sutcliffe* v. *Thackrah* (1974), the Architect would be well advised to follow normal procedures and base each Interim Certificate upon a QS's valuation.
See also *Hedley Byrne & Co. Ltd* v. *Heller & Partners* (1964).

.2.2 Without prejudice to the obligations of the Architect to issue Interim Certificates the Contractor may submit an Application (for payment). The Application shall be in sufficient detail and if the Quantity Surveyor disagrees he shall submit a statement in similar detail identifying the disagreement.

.3 Interim Certificates must be issued at the intervals (usually calendar months) stated in the Appendix until the Certificate of Practical Completion is issued. Partial possession by the Employer will be relevant in this context.
After the issue of the Certificate of Practical Completion, Interim Certificates may be issued as and when further amounts are ascertained as payable to the Contractor from the Employer. Here the *minimum* period between Interim Certificates is one calendar month.

.4 If the Employer fails to pay, the Contractor may give a notice of his intention to suspend performance, and then may suspend until payment occurs. Such a suspension is not a suspension under 27.2.1.1 or a failure to proceed under 27.2.1.2. This clause ensures that JCT 98 complies with the Housing Grants, Construction and Regeneration Act 1996, Section 112.

30.2 The amount stated as due in an Interim Certificate is the gross valuation (Clause 30.2) less the following:

(a)   Retention – as Clause 30.4

(b)   The amount of any advance payment *and*

(c)   The total amount stated as due in previous Interim Certificates (which, normally, will be previous payments).

This provision is subject to any mutual agreement (between the parties) regarding stage, or other, payment. Such agreements are very unusual for JCT – governed building projects but are quite common in international work – see FIDIC Form.

The gross valuation is calculated from the rules in Clauses 30.2.1 and 30.2.2. The amount so calculated must be as at a date not more than 7 days prior to the date of the Interim Certificate.

The components of the gross valuation are:

.1   These are subject to Retention

.1   'the total value of work properly executed by the Contractor' (usually, but not necessarily, calculated by multiplying the measured or assessed quantities of work properly executed by the appropriate rates given in the BQ) including Variations and any applicable formulae adjustments (Clause 13.5 and 40). This also includes items on '*daywork*' but excludes restoration, replacement or repair of loss/damage and disposal of debris which, under Clauses 22B.3.5 and 22C.4.4.2, are treated as if they were a Variation.

Only the value of work which has been *properly executed* in accordance with the Contract must be included; see *Sutcliffe* v. *Chippendale & Edmondson* (1971) and *Townsend* v. *Stone Toms and Partners* (1985). The Architect must decide whether any work has not been properly executed and so must be excluded. If defective work is suspected (say by the QS performing a valuation), the suspected defective work and the value thereof should be drawn to the attention of the Architect for a decision to be made prior to certification.

.2   the total value of 'materials and goods on site' provided that

(a) they are properly for the Works, and

(b) they are not on site prematurely, and

(c) they are adequately protected and stored.

.3 the total value of materials and goods off site, which are 'listed items', subject to the stipulated provisions under Clause 30.3.

.4 the amounts of Nominated Sub-Contractors' items (NSC/C Clause 4.17.1), except final payments.

.5 the amount of Contractor's profit (as Contract Bills or agreement, as applicable) on the amounts included in respect of Nominated Sub-Contractors (clause 30.2.1.4, 30.2.2.5, 30.2.3.2).

Amounts in respect of Nominated Suppliers will be included in materials on or off site or work done and will be subject to additions for Contractor's profit as appropriate.

.2 These are *not* subject to Retention:

.1 amounts to be included due to

(a) Clause 6.2 – fees or charges

(b) Clause 8.3 – inspection and testing

(c) Clause 9.2 – royalties

(d) Clause 21.2.3 – excepted risks

(e) Clauses 22B and 22C – non-insurance against All Risks or Specified Perils by Employer.

.2 (a) loss/expense due to delays (Clause 26.1)

(b) loss/expense due to Antiquities (Clause 34.3)

(c) restoration, replacement or repair of loss/damage and removal and disposal of debris which are treated as a Variation under Clauses 22B.3.5 and 22C.4.4.2.

.3 final payments to Nominated Sub-Contractors – Clause 35.17

.4 payments to the Contractor in respect of 'traditional' fluctuations provisions (Clause 38 and 39)

.5 Nominated Sub-Contractors' amounts as Clause (NSC/C) 4.17.2

.3 *Deductions not* subject to Retention:

.1 (a) Clause 7 – levels and setting out errors by the Contractor

(b) Clause 8.4.2 – work, materials, goods retained but not in accordance with the Contract

(c) Clause 17.2 – defects, shrinkages or other faults not made good under an AI

(d) Clause 17.3 – defects, shrinkages or other faults not made good under an AI

(e) 'traditional' fluctuations as Clause 38 or 39 – amounts to be allowed to the Employer by the Contractor

.2 NSC/C Clause 4.17.3 – Nominated Sub-contractors' 'traditional' fluctuation allowances to the Contractor.

30.3 Materials and goods may be included in Interim Certificates prior to delivery, provided they shall have been listed by the Employer in a list supplied to the Contractor. They must be (collectively referred to here as 'materials'):

.1 & .2 vested in the Contractor so that after payment they shall become the property of the Employer, and if so required the Contractor has provided a bond

.3 are in accordance with the Contract.

.4 (a) set apart from other stocks (i.e. they are 'ascertained')

(b) visibly marked to identify:

(i) the Employer, or the person under whose order the items are being held

(ii) 'their destination as the Works'.

.5 The Contractor provides the Architect with reasonable proof that the materials are properly insured for their full value against 'Specified Perils' from the passing of the property in them to the Contractor to their arrival on site (or adjacent thereto).

Note: Reasonable proof is, normally, the Architect's inspecting the Sub-Contracts and insurance policies and the physical materials.

30.4.1 The rules of Retention deduction by the Employer in any Interim Certificate are:

.1 The percentage is normally 5% or as otherwise specified in the Appendix. Where the Contract Sum is expected to be £500000 or more (at Tender stage) the retention percentage should be not greater than 3%.

Thus, for most major projects, it is reasonable to anticipate that Retention will be 3%; so far, however, the implementation of a recommended low Retention percentage has been rather restricted.

The advocating of a lower Retention percentage for major projects recognises the importance of cash flow in the industry but affords less protection for the Employer against such problems as insolvency of the Contractor. At Practical Completion (of the project or a part thereof), one half of the Retention held is released. In any Interim Certificate following Practical Completion only one half of the Retention Percentage may be deducted and held (Clause 30.4.1.3). Thus, if Retention is stated as 5%, once Practical Completion has been certified only $2\frac{1}{2}$% Retention may be held, the first moiety ($2\frac{1}{2}$%) being released via the Interim Certificate following Practical Completion.

.2   Retention is deducted, as stated, against:

(a) work which has not reached Practical Completion, and

(b) materials and goods (Clauses 30.2.1.2, 30.2.1.3, 30.2.1.4).

.3   Half the Retention Percentage may be deducted in respect of work:

(a) which has reached Practical Completion, *but*

(b) has not achieved a Certificate of Completion of Making Good Defects, *or*

(c) an Interim Certificate to release Retention to a Nominated Sub-Contractor – Final Payment to the NS/C.

.2 'Contractor's Retention' – amounts deducted against Contractor's items.

Nominated Sub-contract Retention' – amounts deducted against Nominated Sub-Contractors' items

Optional Clause 30.4A

Contractors' Bond in Lieu of Retention
There were requests over a period of years for contracts to provide for an optional alternative to retention. JCT discussed with the British Banker's Association and Association of British Insurers the terms of a bond to be given by a Contractor to an Employer in lieu of retention. The bond oper-

ates via an optional clause, which Employers can include in the Contract at 30.4A.

30.4A.2   The provisions (30.2.1 and 30.4) which allow the Employer to deduct and retain a percentage of the total amount included under Clause 30.2.1 and 30.5 shall not apply. However the Architect must instruct the Quantity Surveyor to calculate what the amount would have been; this is in line with the terms of the bond.

30.4A.2   On or before the Date for Possession the Contractor shall provide and maintain a bond in favour of the Employer in the terms set out at Annex 3. The maximum aggregate sum and the expiry date are set out in the Appendix. The executed bond must be supplied to the Employer.

30.4A.3   If the Contractor fails to provide and maintain the bond then retention shall apply from the next Interim Certificate. If the Contractor later provides and maintains the bond then retention shall be released from the next Interim Certificate.

30.4A.4   If at any time the retention that would have been deducted exceeds the amount of the aggregate sum in the bond then the Contractor shall arrange for the aggregate sum to equate to the retention or retention may be deducted.

30.4A.5   If the Employer requires a Performance Bond and a bond in lieu of retention then first recourse shall be to bond in lieu of retention.

30.5   Rules relating to Retention:

   .1   The Employer's interest in any Retentions held is 'fiduciary as trustee'. There is no obligation upon the Employer to invest the monies so held.
       In reality, the Employer probably will invest the funds but any interest earned thereby will accrue to him, not to the Contractor or Nominated-Sub-Contractor, against whom the funds have been retained.

       Re Tout & Finch Ltd (1954) provides a precedent for this contractual provision.

   .2.1 A statement of the Contractor's Retention and Nominated Sub-Contractors' Retention must be prepared by the Architect (or QS under his instruction) on the date of each Interim Certificate.

   .2   The statement must be issued to:

(a) Employer

(b) Contractor

(c) all relevant Nominated Sub-Contractors.

The statement will thus indicate how much Retention is to be deducted in arriving at the amount due. It will typically show the Retention deducted in total and against each Nominated Sub-Contractor.

.3   (Not in LA Edition.) The Employer must:

(a) at the request of the Contractor or Nominated Sub-Contractor, place the Retention held against that party in a separate bank account,

(b) certify to the Architect (copy to the Contractor) the amount so placed

Note: *Rayac Construction Ltd* v. *Lampeter Meat Co. Ltd* (1979), established that retention held by an Employer is held in trust – it is *not* the Employer's money – and so should be set aside as a separate trust fund. This is vital should the Employer go into liquidation – the retention is *not* available to satisfy the Employer's creditors.

Note: Any interest on the sums so placed accrues to the Employer.

.4   If the Employer exercises his rights to withhold and/or deduct against Retention held (as specified by Clause 30.1.1.2), he must inform the Contractor accordingly specifying the amount of set-off by reference to the latest issued statement of Retention.

This Clause is probably a result of the ruling in the case of *Gilbert Ash (Northern) Ltd* v. *Modern Engineering (Bristol) Ltd* (1973), which reversed the decision of the case of *Dawnays Ltd & F.G. Minter Ltd* v. *Trollope & Colls Ltd* (1971).

The *Gilbert Ash* decision related to a main Contractor – Sub-Contractor relationship in which the Sub-Contract contained certain special terms including those relating to payments and set-off. However, the principle laid down by the ruling is applicable to the JCT Form. If one party has a claim against another giving rise to a situation involving contra-charges and set-off, the claim should be quantified as soon as possible, even if approximately, to permit valid set-off. The other party should immediately be given details of the claim and its quantification and afforded the opportunity of agreeing, challenging or somehow reaching a 'compromise' settlement.

Prior to final agreement and settlement of such a claim, any amount set-off should, it is recommended, be lodged with an independent stakeholder. This principle is applicable to main and Sub-Contract relationships.

See also *Mottram Consultants Ltd* v. *Bernard Sunley & Sons Ltd* (1974).

Retention must be released:
Half the relevant holding (whole or part of the Works dependent upon the scope of the Certificate of Practical Completion – whole or part) upon payment of the next Interim Certificate following Practical Completion.

The remainder of the relevant holding of Retention (whole or part) upon payment of the next Interim Certificate following the latter of

(a) expiry of DLP, *or*

(b) issue of the Certificate of Completion of Making Good Defects.

30.6 .1.1 The Contractor must send all documents necessary to properly ascertain any adjustments to be made to the Contract Sum to the Architect (or QS if so instructed by the Architect)

(a) within 6 months of Practical Completion

(b) including documents relating to the accounts of Nominated Sub-Contractors and Nominated Suppliers.

.2 Within 3 months of receipt of the documents from the Contractor (under Clause 30.6.1.1) by the Architect or Quantity Surveyor:
   .1 the Architect (or, if the Architect so instructs, the Quantity Surveyor) must ascertain (unless done already) loss/expense under:

   (a) Clause 26.1 – delays

   (b) Clause 26.4.1 – delays – NS/Cs' work

   (c) Clause 34.3 – Antiquities

   .2 the Quantity Surveyor must prepare a statement of all adjustments to be made to the Contract Sum as per Clause 30.6.2 (except Clause 30.6.1.2.1 – loss/ expense due to delays, antiquities).

The Architect must send a copy of any ascertainment (Clause 30.6.1.2.1) and of the statement of adjustments to the Contract Sum (Clause 30.6.1.2.2) promptly to the Contractor and relevant extracts to NS/Cs.

.2 The Contract Sum must be adjusted as follows:

– the amount of any valuations agreed by the Employer and the Contractor to which 13.4.1.1 refers

– the amounts stated in any 13A Quotation where the Architect has issued a confirmed acceptance

– the amount of any Price Statement accepted pursuant to Clause 13.4.1.2.

From the Contract Sum:
Deduct

.1 (a) all PC sums

(b) all amounts in respect of sub-contractors named under Clause 35.1

(c) certified value of work by a NS/C whose

(i) employment has been determined (Clause 35.24),

(ii) work not in accordance with the sub-contract, and

(iii) Employer has paid, or discharged payment, for the work

(d) all Contractor's profit on (a) to (c)

.2 (a) all provisional sums

(b) value of all Approximate Quantities in the BQ

.3 (a) value of Variations of omissions

(b) value of work as per BQ affected by Variations of omissions, and revalued in consequence (Clause 13.5.5)

.4 (a) any amounts deductible:

(i) Clause 7 – inaccurate levels and setting out retained

(ii) Clause 8.4.2 – retained work, materials or goods not in accordance with the contract

(iii) Clause 17.2 – defects, shrinkages or other faults not made good under an AI

(iv) Clause 17.3 – defects, shrinkages or other faults not made good under an AI

and any amounts allowed or allowable to the Employer under the fluctuations provisions of Clauses 38, 39 and 40, as applicable.

.5 any other amount which the Contract requires to be deducted from the Contract Sum.

Thus Clause 30.6.2.5 may be regarded as a 'longstop' provision – cover omissions of a more minor less frequent nature. One's attention is drawn to the major areas of omission by the preceding 'check-listing' of the relevant areas.

Add

.6 Nominated Sub-Contractors; final accounts in accordance with the NSC/C

.7 final account for the Contractor's work as a Nominated Sub-Contractor (Clause 35.2)

.8 final account for Nominated Suppliers' items (to include for 5% *cash* discount to the Contractor, as Clause 36) *excluding* any VAT input tax to the Contractor. This acknowledges that some items may be subject to VAT which a Contractor may not treat as recoverable input tax

.9 Contractor's profit on Nominated or PC'd items at BQ rates (or as agreed, if appropriate, PC arising from a Provisional Sum Expenditure)

.10 amounts payable to the Contractor by the Employer due to

(a) Clause 6.2 – fees or charges

(b) Clause 8.3 – inspection and testing

(c) Clause 9.2 – royalties

(d) Clause 21.2.3 – premiums paid by Contractor

.11 amount of the valuation of Variations of additions and of Bill work where Variations have caused the conditions to change (Clauses 13.5, 13.5.5), except valuations of omissions under Clause 13.5.2

.12 authorised expenditure against:

(a) Provisional Sums in the BQ

(b) Approximate Quantities in the BQ

.13 claims under:

(a) Clause 26.1 – direct loss/expense due to delays

(b) Clause 34.3 – direct loss/expense due to antiquities

.14 (Not LA Editions) any amounts paid by the Contractor in respect of insurance against All Risks or 'Specified Perils' which should have been effected by the Employer (Clause 22B or C)

.15 fluctuations paid by the Contractor and recoverable from the Employer under the Contract (Clause 38 or 39 or 40)

.16 'Any other amount which is required by this Contract to be added to the Contract Sum'

.17 Any amount to be paid in lieu of ascertainment under 26.1.

.3 The Contractor must be given a copy of the Final Account before the issue of the Final Certificate.

30.7 The Architect must issue an Interim Certificate stating the amounts of all Nominated Sub-Contractors' Final Accounts including the final accounts of NS/Cs in accordance with the provisions of NSC/C. This must be done at least 28 days before the issue of the Final Certificate, even if this means that the one month interval between Interim Certificates is violated.

30.8 The Architect must issue a Final Certificate and inform each Nominated Sub-Contractor of its date of issue. The date of issue will be:
As soon as possible but before the expiry of 2 months from the latest of (in the LA Form reference is made to a period to be stated in the Appendix):

(a) end of the DLP (as per the Appendix), or

(b) date of issue of the Certificate of Completion of Making Good Defects under Clause 17.4, or

(c) date of the Architect's sending to the Contractor:

(i) a copy of the ascertainment under Clause 30.6.1.2.1, and

(ii) a copy of the statement under Clause 30.6.1.2.2.

The Final Certificate must state:

.1   'the sum of the amounts already stated as due in Interim Certificates, plus the amounts of any advance payment.

.2   the Contract Sum adjusted as necessary in accordance with Clause 30.6.2.'

.3   to what the amount relates and the basis of calculation.

the difference between the two sums being expressed as a balance due (either way). From the 28th day after the date of the Final Certificate the balance is a debt due.

30.8 .2   Within 5 days of the issue of the Final Certificate the Employer shall give a written notice detailing the payment to be made, to what the amount of payment relates and the basis on which that amount is calculated (Housing Grants, Construction and Regeneration Act 1996).

.3   The final date for payment shall be 28 days from the date of issue of the Certificate. Within 5 days of the final date for payment the Employer may give the Contractor a notice detailing any amounts to be withheld.

.4   In the absence of a notice under 30.8.3 the Employer may make no deductions.

.5   As in 30.1 failure to pay entitles the Contractor to interest at 5% over Base Rate.

.6   Liability for payment of any balance or interest shall be treated as a debt.

30.9 .1   Subject to the provisions regarding Adjudication or Arbitration, or other proceedings being instigated prior to the issue of the Final Certificate or within 28 days of its issue (and excepting instances involving fraud), the Final Certificate shall, in any proceedings due to the Contract, be conclusive evidence that:

.1   quality of materials and standard of workmanship required to be to the Architect's reasonable satisfaction so comply, and

.2   the Contract Sum has been adjusted appropriately and correctly except for errors of inclusion, exclusion or arithmetic, *and*

.3   all and any extensions of time, due under Clause 25, have been given, *and*

    .4 reimbursement of direct loss/expense to the Contractor under Clause 26.1 (arising out of events under Clause 26.2) is complete and final.

This Clause is not really as meaningless as it appears to be at 'first glance'. Following the case of *Kaye Ltd* v. *Hosier & Dickinson Ltd* (1972) and the wording of the contractual provisions, the Final Certificate is evidence:

(a) of items complying with the Contract where they are required to satisfy the Architect, *and*

(b) innocent and undiscovered mistakes in computations are accepted by the parties.

  .2 If adjudication or arbitration, or other proceedings have been instigated prior to the issue of the Final Certificate, its effect as conclusive evidence is modified:

    .1 by the award of the judgement or settlement, or, *if earlier*

    .2 by the elapsing of a period of 12 months during which neither party has taken any further action in the proceedings, then by any partial settlement reached.

  .3 If adjudication or arbitration, or other proceedings are commenced within 28 days of the issue of the Final Certificate, shall have effect as conclusive evidence save only in respect of all matters to which those proceedings relate.

Colbart v. *Kumar* (1992): (Case concerned IFC 84.) 'Reasonable satisfaction' is not restricted in application to materials, workmanship, etc., expressly reserved by the contract to the opinion of the Architect, but includes all items where approval of such is, inherently, something for the opinion of the Architect. However, where an objective test is specified in the contract documents, that test is the standard required and so the items in operation are removed from those about which the opinion of the architect is the criterion for compliance.

30.10 No certificate of the Architect, except the provisions relating to the Final Certificate, means that any items covered by that Certificate are in accordance with the contract.

Thus, only the Final Certificate provides any evidence of items' compliance with the contractual requirements.

Note: Following *Arenson* v. *Arenson* (1977), if an architect negligently certifies too little, the contractor may sue the Architect.

Clause 31: Statutory tax deduction scheme

See Practice Note 1 (Series 2).

31.1 In this clause the Act means the Income and Corporation Taxes Act 1988;

Authorisation means either a CIS4, CIS5 or CIS6, or a certifying document as prescribed;

construction operations means those operations defined in Section 567 of the Act;

Contractor is a person who is a Contractor for the purposes of the Act and Regulations;

the direct cost of materials means the direct cost to the Contractor of the materials to be used in carrying out the construction operations:

the Regulations means the Income Tax (Sub-Contractors in the Construction Industry) Regulations 1993 SI No. 743 as amended by Income Tax (Sub-Contractors in the Construction Industry) (Amendment) Regulations 1998 SI No. 2622;

statutory deductions means the deductions referred to in the Act;

Sub-Contractor means a person who is a Sub-Contractor for the purposes of the Act;

and *voucher* means either a tax payment voucher in the form CIS25 or a gross payment voucher CIS24.

31.2 The clause is not operative if the Employer is stated not to be a Contractor in the Appendix. If at any time, up to the issue of the Final Certificate, the Employer becomes a Contractor then he shall inform the contractor and the clause becomes operative.

31.3 The Employer shall not make any payment under or pursuant to the Contract unless a valid *Authorisation* has been provided.

31.4 If the Employer is not satisfied with the validity of the *Authorisation* submitted, he shall write to the Contractor notifying him of his grounds for considering the *Authorisation* to be invalid. Once such a notice has been given the Employer shall make no payment until he either receives a valid *Authorisation* or a letter from the tax office confirming validity.

31.5 Where the *Authorisation* is a CIS4 then 7 days before the final payment of any sum due: the Contractor must provide a statement of the *direct cost of materials*; the Employer must make the statutory deductions from the part of the payment which is not for *direct cost of materials*.

31.6   Where the *Authorisation* is a CIS5 or CIS6 or a certifying document then the Employer must make the payment without statutory deduction.

31.7 & 31.8   Provide for changes in *Authorisation*.

31.9   Provides for expiry in *Authorisation*.

31.10   Where *Authorisation* is a CIS4 and the Employer has made payments to the Contractor, the Employer shall next month provide the Contractor with a copy of the CIS25 voucher sent to the Inland Revenue.

31.11   Where *Authorisation* is a CIS6 and the Employer has made payments to the Contractor, the Contractor shall next month provide the Employer with a copy of the CIS24 voucher sent to the Inland Revenue.

31.12   The Employer may correct any errors or omissions in the statutory deductions by repayment or deduction as necessary, subject only to an instruction from the Inland Revenue not to do so.

31.13   Where compliance with this clause involves either party in not complying with any other clause then this clause prevails.

31.14   Where the previous clause is first extended to the resolution of disputes or differences then any other method provided for in the Act or Regulations or any other Act of Parliament or Statutory Instrument is given precedence.

Further Explanation

The statutory tax deductions scheme arose firstly from the Finance (No. 2) Act 1975 and later the Income and Corporation Taxes Act 1988. The scheme was, and continues to be, introduced primarily as a measure to combat tax evasion. The construction industry is a major part of national accounts (often between 8% and 12% of GDP). For many years there has been a problem in construction of the so-called black economy, particularly associated with 'labour only' sub-contracting firms, historically known as The Lump. This scheme makes the operation of tax evasion by 'labour only' firms more difficult because the Employer, under certain conditions, will act on behalf of the Inland Revenue and collect taxes due.

The Appendix will state whether or not the Employer is a 'contractor' or not for the purposes of the Act. The terminology of the Income and Corporation Taxes Act 1988 is very confusing for the construction industry, since the Employer may or may not be a 'contractor' and the Contractor may or may not be a 'Sub-Contractor'.

Clause 32: Number not used

Clause 33: Number not used

Clause 34: Antiquities

34.1 All fossils and antiquities discovered on the site during the execution of the Works are the property of the Employer.

    The Contractor, upon discovering a fossil or antiquity, must immediately:

   .1 use his best endeavours not to disturb the object. He must stop any work likely to disturb or harm the object or impede its removal and inspection. This may obviously delay the project and thereby give grounds for an Extension of Time (Clause 25.4.5.1) and direct loss/expense (but see Clause 34.3.1)

   .2 take steps as thought necessary to preserve the object in the exact location and condition as it was discovered

   .3 inform the Architect or Clerk of Works of the discovery and its exact location.

34.2 The Architect must issue an AI about action concerning the object discovered. Here the Architect is empowered to require the Contractor to allow others onto the site to examine/remove the object. Such a party (probably an archaeologist) is for insurance purposes (Clause 20) considered to be an 'Artist or Tradesman'.

34.3 .1 The Contractor's direct loss/expense due to antiquities is reimbursable at the exercise of the Architect's opinion – the Architect, or QS at his instruction to determine the amount of direct loss/expense.

   .2 If necessary for the calculation of the direct loss/expense incurred, the Architect must state, in writing to the Contractor, any period of Extension of Time awarded due to the discovery of antiquities.

   .3 Any amounts of direct loss/expense are added to the Contract Sum (see above for details).

Conditions: Part 2: Nominated Sub-Contractors and Nominated Suppliers

Clause 35: Nominated Sub-Contractors

In many ways this part of the Conditions of Contract represents also a code of practice. It is thus rather more lengthy and complete than any preceding documents regarding this somewhat complex subject.

35.1 Defines a Nominated Sub-Contractor as one:

where 'the Architect has, whether by the use of a prime cost sum or by naming a sub-contractor, reserved to himself the final selection and approval of the sub-contractor . . .'

The reservation and thence nomination may occur in:

.1 The Contract Bills

.2 An AI (under Clause 13.3) regarding the expenditure of a provisional sum in the BQ

.3 An AI (under Clause 13.2) requiring a Variation where:

.1 it comprises additional work to that in the Contract Drawings or Bills, *and*

.2 Nominated Sub-Contract items of such additional work as are of a similar type to that indicated to be subject to Nomination of a Sub-Contractor in the Contract Bills. (Similar may be read here as, it is suggested, almost identical. This Clause is of obvious importance where the original Sub-Contract work is complete, the additional work constituting, in effect, a new Sub-Contract.)

This provision is a formal recognition of the usual practice in the industry. It does represent a modification of the general principle that only where a PC Sum exists in the BQ or arises due to AIs regarding the expenditure of a provisional sum in the BQ may a nomination occur. This

Clause should make the situation clear, even to the more pedantic.

This represents recognition that it is reasonable for the scope of NS/Cs' work to be extended within their specialist areas if so required by the Variations to the project.

.4   An agreement (not unreasonably withheld) between the Contractor and the Architect on the Employer's behalf. This could include a nomination against the Contractor's measured items in the BQ.

Any Sub-Contractor selected by this procedure must be nominated in accordance with the provisions of this Clause and is termed a Nominated Sub-Contractor.

This Clause applies notwithstanding the provisions of SMM7, Clause A51, which dictates that there must be a PC sum in the BQ for a Nominated Sub-Contractor to be used.

It is usual for a BQ to contain, in relation to each NS/C:

(a) PC Sum (inserted by the QS)

(b) Contractor's Profit thereon (priced by the Contractor)

(c) General Attendances on NS/C (priced by the Contractor)

(d) Other Attendances on NS/C (details by the QS, priced by the Contractor)

(e) Builder's Work in Connection – measured items to be executed, and priced, by the Contractor.

35.2   Contractor tendering for nominated work.

.1   For this situation to arise the following must apply to the work in question:

(a) the Contractor carries out such work directly, in the usual course of his business

(b) the work is included in the Contract Bills and Clause 35 is applicable (see Clause 35.1)

(c) the items of work are set out in the Appendix (by the Contractor)

(d) the Architect is prepared to receive the Contractor's tenders for such items.

The Employer retains the right to reject any tender.

The Contractor may not sub-let any work so obtained without the consent of the Architect.

For the purposes of this Clause, any item against which the Architect intends to nominate an NS/C, arising through AIs regarding expenditure of provisional sums in the BQ (Clause 13.3), is deemed to have been set out in the BQ and included in the Appendix.

The effect is merely that the Contractor may, if the Architect allows, submit a tender for specialist work which he ordinarily executes, whenever such work arises from an AI expending a provisional sum and the Architect wishes to nominate.

.2 Relates to some formal documentation provisions for the purposes of a Contractor's accepted tender in respect of Variations – to refer to Clause 13 wherein references to Contract Drawings and Contract Bills are deemed to refer to the relevant Tender Documents under Clause 35.2.

.3 Only the provisions of Clause 35.2, not all of Clause 35, are applicable to a Contractor whose tender is accepted under this Clause.

Procedure for Nomination of a Sub-Contractor

35.3 Nomination of a Sub-Contractor under Clause 35.1 must be carried out in accordance with Clauses 35.4 to 35.9.

35.4 Lists and identifies documents relating to NS/Cs under the Standard Form of Contract. The usual set will be:

NSC/T   Nominated Sub-Contract Tender, comprising:

              Part 1 – Architect's Invitation to Tender

              Part 2 – Tender

              Part 3 – Particular Conditions (to be agreed by the Contractor and Sub-Contractor nominated under Clause 35.6)

NSC/A   Articles of Agreement (Contractor – NS/C)

NSC/C   Conditions of NSC (incorporated into NSC/A by reference)

NSC/W   Employer – NSC Agreement

NSC/N   Nomination Instruction

35.5   Contractor's right to reasonable objection to a proposed NS/C.

.1   The overriding provision that 'No person against whom the Contractor makes a reasonable objection shall be a Nominated Sub-Contractor'. Such objection to be made in writing within 7 working days of receipt of NSC/N (under Clause 35.6).

.2   Where the Contractor makes a reasonable objection to a nomination, the Architect may:
       (a) issue further instructions to remove the objection (to enable the Contractor to comply with Clause 35.7), or

       (b) cancel the instruction of nomination and issue an instruction either:

           (i) omitting the work (Clause 13.2), or

           (ii) nominating another Sub-Contractor (Clause 35.6).

       The Architect must send a copy of any instructions issued under this clause to the Sub-Contractor.

35.6   The Architect must use NSC/N to issue AIs of nomination. Each must be sent to the Contractor, accompanied by:

.1   (a) NSC/T Part 1 – completed by the Architect,

       (b) NSC/T Part 2 – completed and signed by the NS/C and by (or on behalf of) the Employer as 'approved',

       (c) copy of the numbered tender documents, listed in and enclosed with NSC/T Part 1,

       (d) any additional documents/amendments to them (or to numbered tender documents) which have been approved by the Architect.

.2   NSC/W completed and entered into by the Employer and NS/C.

.3   Confirmation of any changes to the information provided in NSC/T Part 1

       item 7 – obligations/restrictions imposed by the Employer

       item 8 – order of Works: Employer's requirements

       item 9 – type and location of access.

       The Architect must send a copy of the instruction to the NS/C plus a copy of the completed Appendix to the Main Contract.

35.7 On receipt of the instruction, the Contractor must:

    .1 agree with the NS/C and complete NSC/T Part 3 – on such completion, both the Contractor and the NSC must sign that document,

    .2 execute NSC/A with the NS/C and then send a copy of the completed documents (NSC/T Part 3 and NSC/A) to the Architect.

35.8 If, despite using his 'best endeavours', the Contractor has been unable to comply with the requirements of Clause 35.7 within 10 working days of receipt of the instruction from the Architect, he must inform the Architect by written notice that either:

    .1 'the date by which he expects to have complied with Clause 35.7', or

    .2 the non-compliance is due to other matters in the Contractor's notice (e.g. discrepancies/divergencies between documents – as indicated in the footnote, u.1, to the Contract).

    Thus, the Contractor must identify the additional time needed to effect compliance (which should be supported with reasons for the need for extra time) or the reasons why compliance cannot be effected.

35.9 Within a reasonable time of receipt of the Contractor's notice under Clause 35.8, the Architect must:

    .1 where Clause 35.8.1 applies – after consultation with the Contractor, fix a reasonable (in the Architect's opinion) later date for compliance with Clause 35.7;

    .2 where Clause 35.8.2 applies – inform the Contractor in writing, that he considers either:

        (a) the matters identified in the notice do not justify non-compliance with the nomination instruction; then, the Contractor must comply with Clause 35.7 for the nomination in question, or

        (b) the matters identified do justify non-compliance; then the Architect must either:

            (i) issue further AIs to permit the Contractor to comply with Clause 35.7, or

            (ii) cancel the nomination instruction and either:
                – issue an AI omitting the work (Clause 13.2), or
                – nominate another NS/C under Clause 35.6.

The Architect must send a copy of any AI issued under Clause 35.9.2 to the Sub-Contractor.

35.10 Number not used.

35.11 Number not used.

35.12 Number not used.

Payment of a Nominated Sub-Contractor

35.13.1 Upon the issue of each Interim Certificate, the Architect must:

.1 direct the Contractor as to the amounts of any (Interim or Final) Payments to NS/Cs included, as computed by the Architect in accordance with the relevant provisions of NSC/C. This will usually be in the form of a schedule attached to the Contractor's copy of the Interim Certificate.

.2 inform each NS/C of any payments so directed (Clause 35.13.1.1).

.2 The Contractor must make the payments to the NS/Cs in accordance with the relevant NSC/C provisions. (Within 17 days of the date of issue of the Interim Certificate, less 2½% cash discount.)

.3 Prior to the issue of the second and subsequent Interim Certificates, and the Final Certificate, the Contractor must provide the Architect with reasonable proof that payment to any NS/Cs (as required by Clause 35.13.2) has been discharged, usually by production of a receipt detailing the amounts involved. Set-off may form all/part of the due discharge.

.4 If the NS/C fails to give a receipt or other document to enable the Contractor to prove discharge of payment and the Architect is satisfied that the Contractor has paid, the provisions regarding proof of discharge (Clause 35.13.3) are deemed to be satisfied (clause 35.13.5 does not apply); in this case the failure to prove must be the fault of the NS/C in failing to give evidence.

.5.1 If the Contractor fails to provide reasonable proof of discharge of payment to a NS/C (under Clause 35.13.3), the Architect must:

(a) issue a certificate to that effect, stating the amount involved, *and*

(b) issue a copy of that certificate to the appropriate NS/C.

.2 The Employer must pay the amounts due direct to the NS/Cs, in which case:

(a) future payments to the Contractor must be reduced by payments he has failed to discharge to NS/Cs.

(b) VAT due to the NS/Cs in connection with such amounts must also be paid direct

(c) no such payments must be made if the Employer cannot set them off against amounts due from him to the Contractor.

This is always provided that:

(i) the Architect has issued a non-payment certificate, *and*

(ii) the proof failure is not due to the NS/C under Clause 35.13.4

Thus, in summary, direct payment must be implemented where:

– a certificate of non-payment has been implemented

– the proof failure is due to the Contractor's default

– amounts due to the Contractor more than cover any amounts due to NS/Cs after any exercise of set-off rights between the Employer and Contractor for other issues.

.3 Direct payments provision are further modified as follows:

.1 the set-off of the direct NS/C payment by the Employer against amounts due to the Contractor under an Interim Certificate must occur:

(a) at the time of payment to the Contractor, if the set-off leaves any balance to be paid, *or*

(b) within the 14 day period of honouring (see Clause 30) if the set-off leaves no balance for payment to the Contractor.

.2 where the sum due to the Contractor comprises only Retention to be released, the limit of set-off is the amount of Retention to be released to the Contractor (Contractor's retention) in that payment.

.3 where the Employer is to pay 2 or more NS/Cs direct and the amount available, due to the set-off restrictions, is insufficient to meet the payments in full, the Employer must pro rata the amount available (or otherwise fairly apportion the available sum). Any further future amounts must be treated in a similar manner by the Employer to pay NS/Cs direct (e.g. 1st and 2nd releases of Retention).

.4   if the Contractor's business is being wound up by either:

(a) a petition presented to the Court, *or*

(b) a winding up resolution having been passed (unless for amalgamation or reconstruction)

the set-off provisions against the Contractor shall not apply; the relevant time in such instances is if either the first of (a) or (b) has occurred at the time when (under Clause 5.13.5.2) the Employer is to execute the reduction and direct payment to the NS/C.

Note: This Clause will require amendment if the Contractor is not a person to whom the law of insolvency of a company is applicable. In such cases, bankruptcy law applies.

.6   Where the Employer has paid an NS/C, in accordance with Clause 2.2 of Agreement NSC/W, for design work, materials, goods or fabrication which are:

(a) included in the sub-contract sum or the tender sum, and

(b) paid for before the instruction of nomination has been issued

then:

.1   The Employer must send to the Contractor the written statement of the NS/C of the amount (of such items' value(s)) to be credited to the Contractor, and

.2   The Employer may deduct sums from Interim Certificates which include interim or final payments to the NS/C up to the amount of the credit noted in the NS/C's statement. Any such deduction can be made against amounts stated as due to the NS/C only.

35.14 Extension of Period or Periods for Completion of Nominated Sub-Contract Works.

.1   The Contractor's authority to grant Extensions of Time to NS/Cs is limited to the Extensions' being:

(a) in accordance with the provisions of NSC/C, *and*

(b) with the written consent of the Architect.

This applies also to projects to be completed in parts by the NS/Cs.

.2 Clause 2.3 of NSC/C is applicable. The NS/C and the Contractor must send written particulars and estimate of the period involved to the Architect requesting an Extension of Time for the NS/C. The Architect must give his written consent before the Contractor can award any Extension of Time to the NS/C.

35.15 Failure to Complete Nominated Sub-Contract Works.

.1 If any NS/C fails to complete the Sub-Contract Works (or part thereof, if applicable) within the relevant time (Sub-Contract period plus any Extensions), the Contractor should so notify the Architect (copy to NS/C).

Provided that the Architect is satisfied that the Extension of Time provisions have been met (Clause 35.14), he must certify to the Contractor (duplicate to NS/C) that the NS/C has failed to complete by the appropriate date.

.2 The certificate (under Clause 35.15.1) must be issued within 2 months from the date of notification of the NS/C's failure to complete.

35.16 Practical Completion of Nominated Sub-Contract Works.

The Architect must issue a Certificate of Practical Completion of the NS/C Works when such has been achieved *in his opinion* (duplicate to NS/C).

Only one such Certificate is required even if the NS/C Works are completed in sections.

The practical completion is deemed to have occurred on the date of the certificate for the purposes of:

Clause 35.16 – Practical Completion
Clause 35.17, .18, .19 – final payment
Clause 18 – partial possession by Employer.

Westminster City Council v. J. Jarvis & Sons Ltd (1970) (House of Lords) – Jarvis used Peter Lind to execute piling as a Nominated Sub-Contractor; Lind purported to complete their work on the due date. One month later an excavator of Jarvis accidentally knocked a pile which broke off. Tests revealed many piles to be defective due to bad workmanship and/or materials.

In the ensuing claims and legal action over the cost of remedial works, liquidated damages liability and delays, the following was decided:

(a) Lind had achieved Practical Completion of their work on the due date (when they also withdrew from the site).

(b) The piles had latent defects and so the Sub-Contractor was in breach, not delay.

(c) Lind returned to remedy a breach, not to fulfil (i.e. complete) their Contract.

(d) No Extension of Time could be awarded and the Employer therefore had a valid claim against the Contractor (Jarvis) for liquidated damages for delay.

Lind were obviously resolved to bear the liability which the tests revealed, was clearly theirs. They would have to pay either the Employer or the Contractor and so allowed those parties to resolve the legal issue, having executed the necessary remedial work.

As the case was decided, the Contractor was the party to seek redress against them.

It is, perhaps, worth considering an instance where the defects had been discovered prior to Practical Completion of the Sub-Contract Works. Here, it is probable that Practical Completion would have been delayed thereby giving the Contractor grounds for an Extension of Time for the main Contract Works. The Employer could then not claim liquidated damages against the Contractor.

Jarvis John Ltd v. *Rockdale Housing Association* (1986): piling was executed by a NS/C; tests revealed defective piles. Held that a technical breach of the contract resulting from a substantial breach of sub-contract by a NS/C is not 'negligence or default of the Contractor'.

35.17 Early Final Payment of Nominated Sub-Contractors

The Architect may secure final payment to an NS/C (subject to provisions of NSC/C – Clause 5 – being unamended) by including the NS/C's final account in an Interim Certificate issued after Practical Completion of the NS/C Works. This is mandatory if 12 months has elapsed from the date of NS/C's Practical Completion, as certified.

The final payment is subject to provisos:

.1 the NS/C has made good the defects as required contractually, *and*

.2 has sent to the Architect or QS (via the Contractor) all the documents required for the preparation of the final account.

35.18 Upon discharge by the Contractor to the NS/C of the certified final payment (as Clause 35.17):

.1.1 If the NS/C fails to make good any defect as required, the Architect must issue an AI nominating another Sub-Contractor to execute that work.

.2 The Employer must, as far as possible, under NSC/W Recover the sum due to the substituted NS/C from the original NS/C. If this cannot be done the Contractor must pay the appropriate sum to the Employer provided the Contractor had agreed to the substituted NS/C's price prior to that nomination.

.2 Nothing in Clause 35.18 shall override or modify the provisions of Clause 35.21.

The most likely application of Clause 35.18.1.2 is where an NS/C becomes insolvent after completing the Contract work and procuring final payment. Thus, wherever possible it should be ensured that all making good of defects has been completed satisfactorily prior to final payment's being made.

35.19 Despite any final payment to an NS/C:

.1 the Contractor retains full responsibility for loss/damage in respect of items against which final payment to an NS/C has been made to the same extent as for items of uncompleted work, goods and materials up to the date of:

(a) Practical Completion of the Works, *or, if earlier,*

(b) the Employer's taking possession.

.2 The insurance provisions under Clause 22 remain in full force and effect.

Position of Employer in Relation to Nominated Sub-Contractor

35.20 Nothing in the Conditions renders the Employer in any way liable to an NS/C, except as provided in NSC/W.

Clause 2.1 of Agreement NSC/W – Position of Contractor.

35.21 The Contractor is 'not liable' to the Employer in any way regarding NS/C works. The Contractor does, however, retain his full contractual responsibilities regarding the supply of goods, materials and workmanship, including those of an NS/C.

Thus, the Contractor is expressly excluded from any design responsibility which the NS/C may undertake. He does retain responsibilities for the work's execution – i.e. that it complies with the design.

Note: All design is deemed to come to the Contractor from the Employer via his agent, the Architect. The Contractor undertakes no design responsibility.

This Clause regarding the responsibilities of the Contractor operates independently, irrespective of whether the NSC has a responsibility to the Employer under Clause 2.1 of NSC/W.

Restrictions in Contracts of Sale, etc. – Limitation of Liability of Nominated Sub-Contractors

35.22 Any limitation of liability by the NS/C to the Contractor under Clause 1.7 of NSC/C is to be passed on to the Employer.

Note: This may not be possible due to the protection afforded to consumer sales over those afforded to commercial sales by the Unfair Contract Terms Act, 1977 – but see under *Gloucestershire County Council* v. *Richardson* (1969).

35.23 Number not used.

Circumstances where Re-Nomination Necessary

35.24 If, in respect of a Nominated Sub-Contract:

.1   the Architect is of the opinion that the NS/C has made default regarding items as NSC/C, Clauses 29.1.1 to 29.1.4 (grounds for determination) and the Contractor has informed the Architect of the alleged default(s) together with any observations of the NS/C in relation thereto, *or*

.2   the NS/C goes into liquidation, etc. (except for the purposes of reconstruction or amalgamation), *or*

.3   the NS/C determines his employment under Clause 7.7 of NSC/C,

.4   the Contractor has been required to and has determined the employment of the NS/C under Clause 7.3 of NSC/C, *or*

.5   work properly executed, materials or goods properly fixed or supplied by the NS/C have to be re-executed, etc. by the

Contractor or any other NS/C through compliance with an AI (or other power of the Architect) under Clauses 7, 8.4, 17.2 or 17.3

and

the NS/C cannot be required under the terms of the Sub-Contract

and

the NS/C does not agree to carry out the work, etc. to be re-executed

then the following are applicable.

.6 If Clause 35.24.1 applies:

.1 The Architect must issue an AI to the Contractor to give the NS/C notice specifying the default under Clause 7.1.1 of NSC/C. That AI may also instruct the contractor to obtain a further AI prior to determining the employment of the NS/C.

also

.2 Following the giving of the specified notice, the Contractor must inform the Architect if the employment of the NS/C has been determined. If the second AI under Clause 35.24.6.1 has been given, the Contractor must confirm the determination to the Architect.

.3 Upon the Architect being finally informed of the determination, he must re-nominate as necessary. If such determination has been occasioned by the failure of the NS/C to make good defects, etc., the re-nomination must include for that work, etc. to be properly completed by the new NS/C.

.7 .1 If Clause 35.24.2 applies, and the Contractor has an option to determine the employment of the NS/C under Clause 7.2.4 of NSC/C, then Clause 35.24.7.2 applies regarding the written consent of the Architect.

.2 Where the administrator or administrative receiver of the NS/C or the NS/C itself (under voluntary winding up) is:

(a) prepared to continue with the sub-contract,

(b) able to meet the liabilities under the sub-contract, and

(c) both the Architect and the Contractor are reasonably satisfied (with (a) and (b)), the Architect may withhold consent to the determination of the NS/C's employment. However, where (a), (b) and (c) do not apply, the Architect must give written consent to the determination. The Employer and the Contractor may agree otherwise.

.3   Where the employment of the NS/C has been determined:

The Architect must nominate (under Clause 35) a new NS/C to complete the sub-contract work, including all making good, etc. of the work of the original NS/C.

.4   If Clause 35.24.4 applies:

The Architect must nominate (under Clause 35) a new NS/C to complete the sub-contract work, including all making good, etc. of the work of the original NS/C.

.8   .1   If Clause 35.24.3 applies:

The Architect must nominate (under Clause 35) a new NS/C to complete the sub-contract work, including all making good, etc. of the work of the original NS/C.

.2   If Clause 35.24.3 applies:

The Architect must nominate (under Clause 35) a new NS/C to execute the work, etc. to be re-executed under Clause 35.24.3.

Thus, in re-nominating, the Architect must ensure that the new NS/C is willing and instructed to execute all the work, etc. outstanding under the original sub-contract and to complete all necessary making good, etc. of the original NS/C's work. This provision (provided in Clauses 35.24.7.3 to 35.24.8.2) is required because the Contractor cannot be compelled under the contract to execute work which is (or is denoted to be) the subject of a nomination.

.9   Amounts properly payable to new (renominated) NS/Cs must be included in Interim Certificates and added to the Contract Sum.

Except: amounts due to the new NS/C in excess of the price of the original NS/C where the original NS/C validly determined his employment under Clause 7.7 of NSC/C. Such amounts may be deducted by the Employer from sums certified to the Contractor or may be recoverable from the Contractor as a debt (Clauses 35.24.3, 35.24.8.1, 35.24.5 and 35.24.8.2 apply).

.10 The Architect must re-nominate (under Clauses 35.24.6.3, 35.24.7, 35.24.8.1 and 35.24.8.2) within a reasonable time.

Bickerton v. *N.W. Regional Hospital Board* (1969): If the original NS/C 'drops out', the Employer must require the Architect to re-nominate. Subject to the provisions regard execution of specialist work to be the subject of nomination by the Contractor's submitting a tender which is accepted, all items indicated to be executed by an NS/C must, in fact, be executed by an NS/C. The Contractor is not obliged to execute such items.

The costs of the original and subsequent Nominated Sub-Contract items fall on the Employer except any additional costs under Clauses 35.24.5 and 35.24.8.2 which fall upon the Contractor.

Following *Fairclough Building Ltd* v. *Rhuddlan Borough Council* (1985): An Architect's Instruction for re-nomination must include outstanding and remedial items of work; the main contractor has no duty or right to complete the work of the sub-contractor which failed.

The main contractor may reject a re-nomination instruction if the substituted sub-contractor will not undertake to complete the items of work in compliance with the main contractor's programme.

The Architect must re-nominate within a reasonable time of the main contractor's application for an instruction; if not, the delay in re-nominating will be a Relevant Event under Clause 25.4 and may give grounds for a loss and expense claim under Clause 26.

Determination or Determination of Employment of Nominated Sub-Contractor – Architect's Instructions

35.25 The Contractor must not determine any Nominated Sub-Contract without an AI to do so.

.26.1 If the NS/C's employment has been determined under Clauses 7.1 to 7.5 of NSC/C, the Architect must provide the Contractor with the following information and with a direction in an Interim Certificate:

(a) the amount of expenses properly incurred by the Employer, and

(b) the amount of direct loss/damage caused to the Employer by the determination.

The Architect must issue an Interim Certificate which certifies the amount of any monies outstanding to the NS/C (in respect of work, etc. previously uncertified).

.2 If the NS/C's employment has been determined under Clause 7.7 of NSC/C, the Architect must issue an Interim Certificate which certifies the amount of any monies outstanding to the NS/C (in respect of work, etc. previously uncertified).

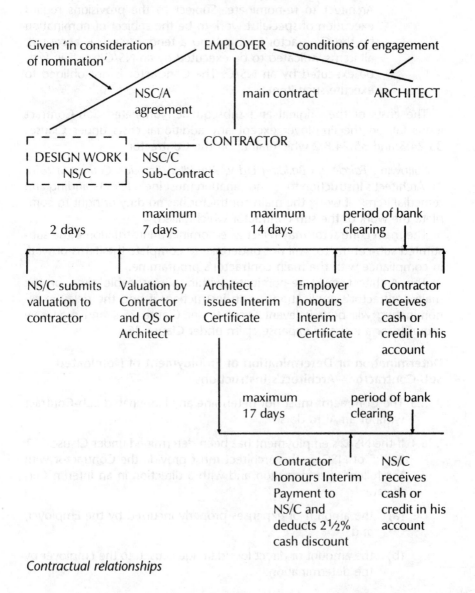

Contractual relationships

Additional noteworthy points regarding NS/Cs

(a)   If the Contractor determines the employment of an NS/C without the requisite AI, he is in breach of Contract.

(b)   If the Architect does not re-nominate as required, the Employer and Contractor should make a separate agreement regarding what is to be done, e.g. use of Artists and Tradesmen; the Contractor to execute the work. Payment will be by an agreed price or *quantum meruit*.

(c)   If an NS/C repudiates the Sub-Contract the Contractor may seek to recover from that NS/C only damages he has incurred directly resulting from the repudiation.

(d)   Any items to be executed by an NS/C are still the responsibility of the Contractor. He must ensure that they are carried out properly by the nominee (or original and subsequent nominees). If the Contractor fails to do this he is in breach of his Contract with the Employer.

Remember:

A Nominated Sub-Contractor is still a Sub-Contractor of the Contractor.

The NS/C is merely nominated by the Architect.

The Contractor is responsible for all work execution.

The Architect is responsible for design (from the Contractor's viewpoint).

The Contractor is entitled to $2\frac{1}{2}$% (1/39th of net accounts to add) cash discount (no other) for prompt payment.

Clause 36: Nominated Suppliers

36.1   .1   Definition of an NSup given.

NSup – nominated by the Architect to supply materials or goods, to be fixed by the Contractor.

Thus, it is usual for the BQ to contain in relation to each NSup:

(a) PC Sum (inserted by QS)

(b) Contractor's Profit thereon (priced by Contractor)

(c) Fix items to be priced by the Contractor.

The nomination by the Architect may occur where:

.1   BQ contains an appropriate PC Sum and the name of the NSup, *or*

BQ contains an appropriate PC Sum and the Architect issues an AI of nomination subsequently (Clause 36.2).

.2   BQ contains a provisional sum. The AI regarding its expenditure gives rise to a PC Sum against which a supplier is nominated in that or a subsequent AI.

.3   BQ contains a provisional sum. The AI regarding its expenditure effectively gives rise to a PC Sum – nominates the supplier if items to be purchased by the Contractor under that AI are available from one source of supply and only one supplier.

.4   Where a Variation (Clause 13.2) requires goods available from a sole source of supply and only one supplier, this is deemed to be a nomination of that supplier against a PC Sum.

Thus the PC Sum against which nomination takes place occurs:

(a)   in the BQ

(b)   by an AI re expenditure of a provisional sum

(c)   as (b) but where there is only one Supplier, not necessarily named as a nomination

(d)   by Variation where there is only one Supplier.

.2 Unless there is a sole Supplier of materials or goods, the supply of the materials or goods must be the subject of a PC Sum in the BQ for the Supplier to be nominated.

This recognises that it has been possible effectively to nominate a Supplier without formally doing so. This is no longer the case. If there is more than one Supplier and more than one source of supply (only one of each now constitutes a nomination), nomination may occur only by an AI regarding the expenditure of a PC Sum (either itself in the BQ or arising out of AIs regarding the expenditure of provisional sum contained in the BQ).

This is clearly demonstrated by:

36.2 The Architect must issue an AI nominating any supplier against any PC Sum (in the BQ or howsoever arising).

36.3 .1 To calculate amounts due in respect of NSup items (Clause 30.6.2.8) only 5% cash discount for prompt payment is allowed to the Contractor (1/19th to add to nett accounts), all other discounts, etc., must be passed on to the Employer.

The Employer must pay the gross amount due (including the cash discount allowance) to the Contractor, which also includes:

.1 any tax or duty, charged under an Act of Parliament, on the goods supplied, not recoverable elsewhere under the Contract (exclusive of VAT input tax which the Contract may reclaim from the Customs and Excise)

.2 the net cost of appropriate packaging, carriage and delivery; subject to any credits for the return of the packaging to the Supplier

.3 fluctuations, subject to the discount provisions as above.

.2 If the Architect believes that the Contractor has incurred (properly) any expense in obtaining the goods etc., which he would not be reimbursed under any other Contract provision, such expense must be added to the Contract Sum.

36.4 The Supplier must enter (or be prepared to enter) into an agreement with the contractor containing the provisions set out below in order that he may be nominated by the Architect. (The only exception is where the Architect and Contractor agree otherwise.)

.1   The quality and standards of items supplied must be as specified or, if appropriate, to the reasonable satisfaction of the Architect.

.2   The NSup must make good or replace (at his own cost) items in which defects appear prior to the expiry of the DLP. The NSup must also bear any direct costs of the Contractor in consequence.

     This is qualified:

    .1   where the items are fixed, a reasonable examination prior to fixing by the Contractor would not have revealed the defects

    .2   the defects are due solely to defective workmanship or materials in the items supplied. It is invalidated if the defects were caused by someone outside the control of the NSup (e.g. inadequate storage by the Contractor).

.3   Delivery must be in accordance with a programme agreed between the Contractor and NSup, *or* in accordance with the Contractor's reasonable directions.

     Any programme for delivery agreed between the Contractor and the NSup may be altered due to:

    (a) force majeure,

    (b) civil commotion, strike, lockout, etc.,

    (c) AIs regarding Variations (Clause 13.2) or provisional sums (Clause 13.3),

    (d) Failure of the Architect to supply the NSup with necessary information for which the NSup had applied in writing at an appropriate time (reasonably close to the date on which the information was required),

    (e) exceptionally adverse weather.

.4   The NSup must allow the Contractor 5% cash discount for payments made for items within 30 days from the end of the month in which those items were delivered (as discussed earlier, any other discounts must be passed on to the Employer).

.5   The NSup need not deliver any items after a determination of the Contractor's employment *except* any for which he has been properly paid.

.6   The operation of Clause 36.4.4 fully discharges the payment provisions.

.7   Ownership of the items passes to the Contractor upon their delivery, even if they have not been paid for. (However, see *Aluminium Industrie Vaassen BV* v. *Romalpa Aluminium Ltd* (1976) where a retention of title clause applied.)

.8   If any dispute or difference between the Contractor and the nominated Sub-Contractor is referred to Arbitration then 41B applies (the Arbitration Provisions).

.9   No Contract of sale conditions may prevail over the conditions set out in this Clause.

Note: Use of JCT Standard Form of Tender by Nominated Suppliers (TNS/1) is advisable.

36.5 .1   If any liability restriction or exclusion (as Clauses 36.5.2 or 36.5.3) exists between the NSup and Contractor and has the written approval of the Architect, the Contractor may restrict his liability to the Employer to the same extent.

.2   The Contractor must have the Architect's written approval of any restrictions of the NSup's liability prior to entering into a contract with that NSup.

.3   All nominations of Suppliers by the Architect must comply with Clause 36.4.

Remember   the provision of the Unfair Contract Terms Act, 1977 regarding liability restrictions – exclusion or exemption clauses, particularly related to 'consumer sales'.

Note: Exemption clauses, even if complying with requirements of notice and reasonableness, normally are constructed by the courts strictly against the parties relying on them.

Ailsa Craig Fishing Co. Ltd. v. *Malvern Fishing Co. Ltd. & Another* (1983), differentiates clauses excluding liabilities from clauses limiting liability – the latter are viewed with less 'hostility' by the courts.

If an NSup wishes to restrict his liabilities or obligations under a proposed supply Contract, the Contract must obtain the Architect's written consent to the restrictions or else the Contractor himself assumes the liabilities.

Note: Unlike the provisions relating to NS/Cs, the contractor has
no right of 'reasonable objection' to an NSup.

Gloucestershire County Council v. *Richardson* (1969) (House of
Lords): The ruling in this case that, due to nomination, the
Supplier's exclusion clauses were well known to the Architect
(and, by implications of agency, the Employer) who then
nominated and, thereby, gave the Contractor no option but
to go to that Supplier, the implied warranty regarding fitness
for purposes and latent and patent defects was also excluded
from the Contract between the Employer and Contractor in
this regard. The exclusion clause was validly passed on.

As indicated above regarding terms of the contract of sale,
the Standard Form now stipulates what must be contained in
a contract of sale and how liability exclusions are to be dealt
with and possibly passed on. The fact of nomination by the
Employer's expert agent, the Architect, may be sufficient to
overcome the problems of passing on such liability through
the operation of the Unfair Contract Terms Act, 1977.

Conditions: Part 3: Fluctuations

Clause 37: Choice of fluctuation provisions – entry in Appendix

37.1 The chosen fluctuation provisions applicable to the Contract must be stated in the Appendix.

The choice is:

Clause 38 – Contribution, Levy and Tax Fluctuations
or
Clause 39 – Labour and Materials Cost and Tax Fluctuations
or
Clause 40 – Use of Price Adjustment Formulae

37.2 If neither Clause 39 nor 40 is identified in the Appendix, Clause 38 applies.

Thus, unless the Appendix provides for the Contract to be executed on the basis of 'full' fluctuations, 'partial' fluctuations provisions are applicable (often termed firm price Contract).

The three fluctuations Clauses are contained in a separate booklet detailing the various versions of each Clause applicable to the Private Edition (separate booklet for LA Edition).

No version of Clause 38 exists for the Approximate Quantities Contract and no version of Clause 40 exists for the Without Quantities Contract.

37.3 Where the Architect has accepted a 13A Quotation, such work is not subject to the fluctuations' Clauses 38, 39 and 40.

Clause 38: Contribution, levy and tax fluctuations

Clause 38 is not suitable for use with the Approximate Quantities Edition.

38.1 The Contract Sum is deemed to have been calculated as detailed by this Clause (whether it is or not is a matter for decision by the Contractor at the tendering stage).

Note: at this point it is useful to consider some definitions as contained in Clause 38.6 (also 39.7).

38.6.1 (39.7.1) 'Date of Tender': the date 10 days before the date fixed for the receipt of Tenders by the Employer.

38.6.2 (39.7.2) 'Materials' and 'Goods': exclude consumable stores, plant and machinery, but include:

   (a) timber used in formwork

   (b) electricity

   (c) fuels, where specifically so stated in the Contract Bills – see Clause 38.2 (39.3).

38.6.3 (39.7.3) 'Workpeople': persons whose rates of wages, etc., are governed by the National Joint Council for the building Industry or some other wage-fixing body for trades associated with the building industry.

38.6.4 (39.7.4) 'Wage-fixing Body': a body which lays down recognised terms and conditions of workers. The 'recognised terms and conditions' concerns:

   – trade or industry of the employer (or a section thereof),

   – workers in comparable employment in that trade/industry are 'governed' by the terms and conditions,

   – settlements by agreement or award, the parties to which are employers' associations and independent trade unions which represent substantial proportions of relevant employers and workers.

The definitions are applicable to Clause 38 (39) only.

.1   The prices in the Contract Bills are based upon the types ('tender type') and rates ('tender rate') of contribution, levy and tax payable by a person in his capacity as an employer and which are payable by the Contractor at the Date of Tender.

.2   If any tender types or tender rates change, are deleted or new ones are introduced after the Date of Tender, the net alteration must be paid or allowed by the Contractor.

This applies in respect of:

.1   workpeople on site, and

.2   the Contractor's workpeople off site but working upon or in connection with the Contract (e.g. production of goods – joinery, etc.)

Note: both cases – the operatives are the Contractor's direct employees only.

Note: levies, etc., under the Industrial Training Act, 1964 (CITB levy and payment) are expressly excluded.

.3   Other direct employees of the Contractor working on or in connection with the Works are, for the purposes of this Clause, created as craft operatives (as prescribed by Clause 38.1.4).

.4   In such instances, however, the following provisos apply.

(a) Each employee must have worked on or in connection with the Contract for at least two working days during the week against which the claim is applicable. Time aggregation is allowed in respect of whole working days only.

(b) The highest properly fixed craft operative's rate must be used, provided such a craft is employed by the Contractor (or Domestic S/C).

(c) The Clause is applicable to those employed by the Contractor as defined by the Income Tax (Employment) Regulations, 1973, (the PAYE Regulations) under S204 of the Income and Corporation Taxes Act 1970.

.5   'Tender type' and 'tender rate' are redefined for Clause 38.1.6 in respect of refunds to the Contractor in his role as an employer of labour. The definition is extended to include premiums receivable by an employer of labour. Prices in the

BQ are based on tender types and rates as at the Date of Tender.

.6   The net alteration to any tender types or rates from those applicable at the Date of Tender must be paid to or allowed by the Contractor.

.7   Premiums are defined as (see Clauses 39.1.5 and 38.1.6):

'. . . all payments howsoever they are described which are made under or by virtue of an Act of Parliament to a person in his capacity as an employer and which affect the cost to an employer of having persons in his employment.'

.8   Any direct operatives who are 'contracted out' (Social Security Pensions Act, 1975) are deemed not to be so for the purposes of calculating fluctuations in respect of employers' contributions.

Thus, no employee may be contracted out for the calculation of employers' contributions (NI contributions).

.9   Contributions, levies and taxes are defined for the purposes of this Clause in a very similar manner to the above definition of premiums – what an employer must pay under Acts of Parliament as an employer of labour.

Note: This will include statutory insurances against personal injury and death.

38.2   The Contract sum is deemed to have been calculated in the prescribed manner, in respect of materials, goods and fuels, and is subject to adjustments as detailed.

.1   Prices in the Contract Bills are based upon duty and tax types ('tender type') and rates ('tender rate') applicable at the Date of Tender on import, sale, purchase, appropriation, processing or use (except VAT reclaimable as the Contractor's input tax) on:

(a) materials

(b) goods

(c) electricity

(d) fuels – if so specifically stated in the BQ and specified on a list completed by the Contractor and attached to the BQ.

This is provided the duty or tax is applicable under an Act of Parliament.

.2   If any tender type or rate is altered from that applicable at the Date of Tender (including electricity and, if applicable, fuel for the temporary site installations) or a deletion or new type or rate occurs, the net difference between that paid by the Contractor and what he otherwise would have paid must be paid to or allowed by him (except, of course, VAT input tax).

38.3   Domestic Sub-Contractors' items.

.1   The Domestic Sub-Contract must incorporate provisions to the same effect as those of Clause 38 of the main Contract together with any appropriate percentage (Clause 38.7).

.2   Any adjustment to the Domestic Sub-Contract sum due to the operation of the Clause 38.3.1 must be passed on to the Employer by the Contractor.

38.4 .1   The Contractor must give written notice to the Architect of the occurrence of any events regarding the provisions of:

.1   Clause 38.1.2 – tender types and rates and workpeople

.2   Clause 38.1.6 – tender types and rates and refunds

.3   Clause 38.2.2 – tender types and rates and materials, etc.

.4   Clause 38.3.2 – Domestic Sub-Contractors' fluctuations.

.2   The written notice is a condition precedent to any payment to the Contractor under this Clause and so must be given within a reasonable time from the occurrence of the event to which it relates.

Thus, a notice is required each time a fluctuation event occurs, although one notice may properly cover several events which occurred at or about the same time.

.3   The QS and Contractor may agree the net amount of each fluctuation due to each notified event.

.4   Any amount of fluctuations shall form an adjustment to:

.1   the Contract Sum, *and*

.2   any determination payments (Clause 28.2.2.1 and 28.2.2.2). This is subject to the provisions of Clauses 38.4.5, 38.4.6 and 38.4.7.

.5   The Contractor must submit to the QS or Architect any evidence they may reasonably require to calculate the amounts of fluctuations. This must be done as soon as reasonably practical. Where a fluctuation is claimed in respect of employees other than 'workpeople' (not operatives – Clause 38.1.3), the evidence must include a certificate signed by or on behalf of the Contractor each week certifying the validity of the evidence. (This is applicable also to Domestic Sub-Contractors' fluctuations.)

.6   Fluctuations adjustments must never alter the amount of Contractor's profit included in the Contract Sum. Thus, fluctuations must be paid net.

This seems to suggest the sum constituting the Contractor's profit must not be altered by fluctuations adjustments (as the percentage on cost would be).

.7   Fluctuations are not adjustable in respect of payments to the Contractor after the Completion Date, as appropriately amended – the fluctuations are 'frozen' at the Completion Date.

Note: This amends the ruling of Salmon ⊔ in *Peak Construction (Liverpool) Ltd* v. *McKinney Foundations Ltd* (1971).

.8   Clause 38.4.7 applies only if:

.1   Clause 25 applies unamended (Extension of Time), *and*

.2   the Architect has made an award in respect of ever written Extension of Time notification (Clause 25).

38.5   Fluctuation provisions are not applicable in respect of:

.1   Dayworks (Clause 13.5.4)

.2   NS/Cs and Nominated Suppliers (but are included in the NSC or Contract of Sale)

.3   Work for which the Contractor's tender under Clause 35.2 has been accepted – Contractor acting as an NS/C also. (Clause 35 tender conditions to apply.)

.4   VAT changes.

38.6   Definitions as detailed above – see under clause 38.1

38.7   The percentage stated in the Appendix must be added to fluctuations paid to or allowed by the Contractor under Clauses:

.1   38.1.2

.2   38.1.3

.3   38.1.6

.4   38.2.2

Note: Following *J. Murphy Ltd* v. *London Borough of Southwark* (1981), the term 'workpeople' in the fluctuations clauses does not include labour only sub-contractors.

Clause 39: Labour and materials cost and tax fluctuations

As this is the full, traditional fluctuations clause, much of the Clause 39 provisions are reproduced – see Clauses 38.6 and 39.7 – the definitions. Again, there are several distinct sections.

Clause 39.1 – wages
Clause 39.2 – labour taxes
Clause 39.3 – materials prices (including taxes)
Clause 39.4 – Domestic Sub-Contractors
Clause 39.5 – fluctuations calculations
Clause 39.6 – work not applicable
Clause 39.7 – definitions

39.1 The Contract Sum is deemed to have been calculated as detailed by this Clause and is subject to adjustment as specified.

.1 The prices in the Contract Bills are based upon rates of wages and the other emoluments (payments) and expenses which will be payable by the Contractor. This includes:

(a) Employer's liability insurance (contractor as an employer of labour), and

(b) Third party insurance, and

(c) Holiday credits.

The payments are in respect of:

.1 Workpeople on site, *and*

.2 Contractor's workpeople off site but working on or in connection with the Contract.

The payments must be in accordance with:

.3 the rules or decisions of the NJCBI, or other appropriate wage-fixing body, as applicable to the Works and which have been promulgated at the Date of Tender.

Note: promulgated – published as coming into force or having authority. Thus, any wage changes which have been taken into account in the calculation of the Contract Sum are, therefore, not subject to the fluctuations provisions.

.4   any incentive scheme/productivity agreement as advised and recognised by NJCBI (Working Rule Agreement 1.16, or its successor, including the general principles provisions of the Working Rule Agreement) or other appropriate body.

.5   the terms of the Building and Civil Engineering Annual and Public Holiday Agreements (of NJCBI or other appropriate body) as applicable to the Works and which have been promulgated at the Date of Tender.

The prices in the BQ are deemed to be based also upon the rates or amount of any contribution, levy or tax payable by the Contractor as an employer in respect of (or calculated by reference to):

(a) rates of wages, and

(b) other emoluments, and

(c) expenses, including holiday credits as specified in this Clause.

Note: By the provision of Clause 39.2.2, the contributions and levies of the CITB, etc. are expressly excluded from the fluctuations provisions.

.2   If any rules, decisions or agreements promulgated after the Date of Tender alter the rates of wages, other emoluments and expenses, the net amount of the change must be paid to or allowed by the Contractor.

Such adjustment is inclusive of any consequential change in the cost of:

(a) Employer's liability insurance, and

(b) third party insurance, and

(c) any contribution, levy or tax payable by a person as an employer of labour.

Thus, if wage rates increase after the Date of Tender due to a promulgation after that date, both that increase and any consequential increase in, say, employer's liability insurance are recoverable by the Contractor. (Employer's liability insurance is often based upon the wage bill of a firm.)

However, if the Employer's liability insurance premium is increased due to (a) a wage increase and (b) inflation generally, then only (a) is recoverable under this Clause.

To be recoverable, the consequential increase must be directly and solely due to one or more of the specified causes (increase in rates of wages on the Contract), not just more expensive insurance.

Despite the House of Lords ruling under the 1963 Form in *William Sindall Ltd* v. *N.W. Thames Regional Health Authority* (1977), it would now appear from the wording of this Clause 39 that increases in the cost of productivity bonuses directly resulting from increases in standard rates of wages (established by NJCBI or other appropriate wage-fixing body) are recoverable. The applicable proviso is that the bonus scheme is:

(a) covered by the Working Rule Agreement, as required by Clause 39.1.1.4, *and*

(b) recognised by the unions.

It is suggested that a casual, unrecognised, site agreement would be outside the scope of this Clause and so the Sindall ruling would apply:

(a) the consequential bonus increase is not recoverable, but

(b) if a productivity or other bonus scheme is being operated 'casually', although consequential increases in these payments are not recoverable, increases in guaranteed minimum bonus are recoverable – they represent guaranteed minimum payments.

.3 Contractor's employees, outside the definition of workpeople, engaged on or in connection with the Works, are subject to fluctuations recovery as if they were craft operatives.

.4 (This is as Clause 38.1.4 but is used here to cover wage rate increases.)

Note: The wording of the Clause implies that production bonus, other than guaranteed minimum, payments should be ignored when calculating fluctuations in respect of these employees.

.5 Fares, etc.

The prices contained in the BQ are based upon:

(a) transport charges incurred by the Contractor (as set out in the basic transport charges list attached by the Contractor to the BQ) in respect of workpeople – Clauses 39.1.1.1 and 39.1.1.2, *or*

DAWSON

ADVANCE ANNOUNCEMENT OF NEW BOOK

Aitken, S.
GEOGRAPHIES OF YOUNG PEOPLE the morally contested spaces of identity.

224 pp.
Aug 01, Routledge.

ISBN/ISSN	Price	Format
0415223946	£60.00	Cloth
0415223954	£18.99	Paper

SUMMARY This work traces some of the changing scientific and societal notions of what it is to be a young person, and argues that there is a need to rethink how we view childhood spaces, child development and the politics of growing up.
(Critical geographies)

Order From Dawson UK Ltd, Book Division,
Crane Close, Wellingborough, Northants NN8 2QG, UK.
Tel: 01933 274444 Fax 01933 225993

CLASSIFICATION
300
911.3
READERSHIP LEVEL
Undergraduate

Title No.
291220

Registration No.
10484

692.8 FEL

(b) reimbursement of fares to workpeople (Clauses 39.1.1.1 and 39.1.1.2) under the rules of NJCBI (or other appropriate body) promulgated at the Date of Tender.

.6 The net amount of any changes in cost of fares, etc., to the Contractor, as defined above, must be paid to or allowed by him on the basis of:

.1 changes in transport charges set out in the basic list of transport charges occurring after the Date of Tender

.2 actual reimbursement taking effect or promulgation of changes (as appropriate) after the Date of Tender.

39.2 Largely a reproduction of parts of Clause 38.

.1 Clause 38.1.1

.2 Effectively Clause 38.1.2

.3 Clause 38.1.3 incorporating Clauses 38.1.4 and 39.1.4

.4 Clause 38.1.5

.5 Clause 38.1.6

.6 Clause 38.1.7

.7 Clause 38.1.8 except where contracted out employee pension schemes are under the rules of NJCBI, or other appropriate body, such that contributions are covered by the provision of Clause 39.1 and are thereby changes in contributions and recoverable.

.8 Clause 38.1.9

39.3 Materials – The Contract Sum is deemed to have been calculated as specified and is subject to adjustment accordingly.

.1 The prices contained in the BQ are based upon the market prices of materials, goods, electricity and, where specifically so stated in the BQ, fuel-list required (as Clause 38.2.1). The market prices must have been current at the Date of Tender and are termed 'basic prices'. The prices set out by the Contractor in the list (basic price list) and attached to the BQ are deemed to be the basic prices of the items specified therein.

Note: Electricity and any fuels must be consumed on site for the execution of the Works, including temporary site installations.

.2   Any changes in the market prices of items specified in the basic list (including electricity and fuels for the Works and temporary site installations) occurring after the Date of Tender are the subject of fluctuations adjustments of the net changes.

.3   Market prices include any tax or duty (except VAT input tax of the Contractor) on the import, purchase, sale, appropriation, processing or use of goods, materials, electricity or fuels as specified, provided such tax or duty is governed by an Act of Parliament.

'Market Price' – Lord Kinnear in *Charrington & Co. Ltd* v. *Wooder* (1914):

'I am unable to accept the contention that the term 'market price' has a fixed and definite meaning which must attach to it invariably, in whatever contract it may occur, irrespective of the context or the surrounding circumstances. The argument was rested chiefly on the force which, it is said, must be given to the word 'market'. In a different connection this may be a technical term, but the covenant in question is not used in any technical sense, and in ordinary language it is a common word, of the most general import. It may mean a place set apart from trading, it may mean simply purchase and sale, there are innumerable markets each with its own customs and conditions. Words of this kind must vary in their signification with the particular objects to which the language is directed; and it follows that a contract about a market price cannot be correctly interpreted or applied without reference to the facts to which the contract relates.'

Thus, market price must be viewed in context. If there is only one supplier, that supplier's price is the market price.

If there are many suppliers the market price will have a range, the Contractor's usual supplier's price indicating the market price for that Contractor.

If an item becomes restricted in the suppliers from whom it is obtainable, the Contractor must obtain that item from one of these suppliers, possibly at an elevated price but that is the market price of the item to the Contractor at that time.

Usually, Contractors obtain commodities from the suppliers whose quotations were used in the preparation of the basic price list.

39.4 Work let to Domestic Sub-Contractors – corresponding provisions to those of Clause 38.3 but of increased scope to cover wage fluctuations.

39.5   As Clause 38.4 but the notice to be in respect of

    .1   .1   Clause 39.1.2 – wages

          .2   Clause 39.1.6 – transport and fairs

          .3   Clause 39.2.2 –tender types and tender rates – alterations

          .4   Clause 39.2.5 –tender types and tender rates – alterations/deletions/additions

          .5   Clause 39.3.2 – market prices

          .6   Clause 39.4.2 – Domestic Sub-Contractors

    .2   Clause 38.4.2

    .3   Clause 38.4.3

    .4   Clause 38.4.4

    .5   Clause 38.4.5

    .6   Clause 38.4.6

    .7   Clause 38.4.7

    .8   Clause 38.4.8

39.6   Work not subject to fluctuations as Clause 38.5.

39.7   Definitions – as Clause 38.6.

39.8   .1   Percentage addition to be added, as stated in the Appendix to fluctuations under:

        .1   Clause 39.1.2

        .2   Clause 39.1.3

        .3   Clause 39.1.6

        .4   Clause 39.2.2

        .5   Clause 39.2.5

        .6   Clause 39.3.2.

In respect of the provisions of Clause 39.1.5, transport charges, and 39.3.1, materials including electricity and fuels, it would be of value if QSs were to provide suitably headed sheets on which the Contractor should enter his items to comprise the basic lists for fluctuations recovery and to submit these with his tender or Bills, as required.

Clause 40: Use of price adjustment formulae

The formula method of fluctuations and adjustments is rapidly becoming more widely accepted and used. It provides for full fluctuations recovery and, often, is termed the NEDO Formula, after the indices upon which it is based. There is no provision for the use of formulae fluctuations in the Without Quantities Editions.

The adjustment applies to all payments *except* releases of Retention as these will have been the subject of formula adjustments already.

Note: Clause 30.2.1.1 provides for Retention to be deducted from formula fluctuations adjustments.

40.1 .1.1 Prescribes for the Contract Sum to be adjusted in accordance with Clause 40 and the JCT's Formula Rules current at the Date of Tender.

    .2 Any formula adjustment is exclusive of VAT and must not affect the operation of Clause 15 – VAT Supplemental Provisions and the VAT Agreement.

    .2 The Definitions in Rule 3 of the Formula Rules are to apply to Clause 40.

    .3 The adjustments are to be effected in all certificates for payment. (LA Forms require any non-adjustable element to be deducted prior to the adjustment being made.)

    .4 If any adjustment correction is required (Rule 5), it must be effected in the next certificate for payment.

40.2 Interim Valuations *must* be made before the issue of each Interim Certificate. Clause 30.1.2 is deemed amended accordingly.

40.3 Articles manufactured outside the UK (Rule 4(ii)) – Contractor to append a list to the BQ giving the sterling market price of each item (delivered to site) as at the Date of Tender.

Any variation in the market price is subject to adjustment under provisions similar to those of Clause 39, i.e. exclusive of contractor's input VAT.

40.5 The QS and Contractor may agree any alteration to the methods and procedures for the formula fluctuations recovery. Any amounts of adjustments so determined are deemed to be the formula adjustment amounts.

This is subject to:

.1 the amounts of adjustment calculated after amendment must approximate to those obtained by operation of the Rules as laid down, *and*

.2 any such amendments must not affect the provisions relating to Sub-Contractors (as Clause 40.4).

40.6 .1 Prior to the issue of the Final Certificate, if the publication of the Monthly Bulletins is delayed or ceases, thereby preventing the formula adjustment being properly effected, adjustment is to be made on a fair and reasonable basis to determine the amount due in each Interim Certificate so affected.

.2 Prior to the issue of the Final Certificate, if the publication of the Monthly Bulletins is recommenced, Clause 40 and the Formula Rules are to apply retrospectively. The index adjustment from the Bulletins must be carried out and will prevail over the fair adjustments made.

.3 During any cessation of publication of the Monthly Bulletins, the Contractor and Employer must operate the relevant provisions of Clause 40 and the Formula Rules so that on publication being recommenced the appropriate formula adjustments may be made retrospectively.

40.7 .1 .1 If the Contractor does not complete the Works by the Completion Date, as applicable, the value of work completed after the date is subject to formula adjustment on the basis of the indices applicable at the relevant Completion Date.
Thus, items completed in a time over-run are adjusted under the formula method as if they were completed in the month prior to the applicable Completion Date.

.2 If items completed in an over-run period are adjusted in any other way, such adjustment must be corrected to accord with that specified by Clause 40.7.1.1, as above.

.2 Clause 40.7.1 does *not* apply *unless*:

.1 Clause 25 (Extension of Time) applies, unamended, to the Contract, *and*

.2 the Architect has fixed the Completion Date as appropriate, in writing, in respect of every written notification by the Contractor under Clause 25.

Conditions: Part 4: Settlement of Disputes – Adjudication – Arbitration – Legal Proceedings

Clause 41A: Adjudication

41A.1 This clause applies where either party refers a dispute to Adjudication pursuant to Article 5.

41A.2 The Adjudicator can be either an individual agreed by the parties or an individual nominated by the 'nominator' in the Appendix. An Adjudicator must execute the JCT Adjudication Agreement; agreement by the parties must be within 7 days of the referral or the nomination must take place within 7 days. All this ensures that the appointment and the terms of the Adjudication comply with the provisions of the Housing Grants, Construction and Regeneration Act 1996.

41A.3 If the Adjudicator dies or becomes otherwise incapacitated, the parties agree on a replacement or apply for a further nomination in the same terms.

41A.4 When a Party requires a dispute or difference to be referred to Adjudication then that Party shall give notice to the other Party of the intention and briefly describe the dispute or difference in the notice. If an Adjudicator is agreed or appointed within seven days of the notice then the Party giving the notice shall refer the dispute or difference to the Adjudicator within seven days of the notice. If the Adjudicator is not agreed or appointed within seven days of the notice the referral shall be made immediately on such agreement or appointment. Again the Party shall include with the referral particulars of the dispute or the difference together with a summary of contentions relied upon, the statements of the relief or remedy which is sought and any material the Adjudicator is required to consider. The referral is copied simultaneously to the other Party.

The referral and documentation may be provided by: actual delivery, Fax, special delivery or recorded delivery. If the delivery is by Fax then copies must be provided by first class post or actual delivery. Without proof to the contrary, delivery is deemed to be 48 hours after posting (excluding Sunday and public holidays).

41A.5 This clause gives detailed guidance on the conduct of the Adjudication; the guidance is complementary to the JCT Adjudication Agreement. Comments are made here and in a Further explanation later.

Immediately upon receipt of the referral and its accompanying documentation the Adjudicator must confirm the date of receipt to the Parties. The Party not making the referral, (in arbitration or litigation they would be called the Respondent) may in a similar manner to the party making referral (the Claimant) send to the Adjudicator within 7 days of the day of the referral a written statement of the contentions on which he relies and any materially he wishes the Adjudicator to consider. Naturally this response is to be copied to the party making the referral.

Within 28 days of the referral the Adjudicator (acting as an Adjudicator for the purposes of section 108 of Housing Grants, Construction and Regeneration Act 1996 and not as an expert or an arbitrator) makes a decision and sends this immediately in writing to the Parties. The Party making a referral may consent to an extension of 14 days, and if agreed between the Parties a longer period, within which to reach a decision.

The Adjudicator is not obliged to give reasons for decisions; this is important since without reasons there is very little to appeal against.

In reaching a decision the Adjudicator would act impartially and set his own procedure. At his absolute discretion be may take the initiative in ascertaining the facts and the law, including the following:

Application of his own knowledge and or expertise.
The opening-up, reviewing and revising of any certificate, opinion, decision, requirement or notice issued, given or made under the Contract as if no such certificate, opinion, decision, requirement or notice had been issued, given or made.
Requiring from the Parties any further information.
The carrying-out of further tests.
Visiting the site or any other place where work has been carried out.
Obtaining further information as he considers necessary from Parties, Employees or representative, provided prior notice is given.
Obtaining from others such information and advice as he considers necessary on technical and legal matters, provided prior notice is given, and also an estimate of costs.

Having regard to any term of the contract relating to the payment of interest, deciding the circumstances in which or the period for which a simple rate of interest should be paid.

Any failure by either party to enter into the JCT Adjudication Agreement or to comply with the requirements of the Adjudicator will not invalidate the decision of the Adjudicator.

The parties will meet their own costs of the Adjudication except that the Adjudicator may direct as to who should pay the cost of any test or opening-up, if required. This is an important clause in terms of dispute resolution, since at law, and in arbitration, the so-called English Rule requires that generally a loser in any action pays the winner's costs. The provision here against the English Rule may prevent the larger and more powerful party from intimidating an opponent by the expenditure of large amounts on legal advice, with the smaller and weaker party being at risk of liability for the costs should they lose.

41A.6 The Adjudicator may decide how payment of his fees and expenses are to be apportioned between the parties. This gives the Adjudicator the power to award his costs and fees against a party. If the Adjudicator fails to make an award on his costs, the parties will bear those costs in equal proportion.

The Parties shall be jointly and severally liable to the Adjudicator for his fee and all expenses reasonably incurred. This means for the Adjudicator, if one party is unable to pay their proportion of his fee the other party is liable for that amount.

41A.7 The decision of the Adjudicator is binding on the Parties until the dispute or difference is finally determined by arbitration or by legal proceedings or by agreement. The arbitration or legal proceedings are not an appeal against the decision of the Adjudicator but are a consideration of the dispute or difference as if no decision had been made by an Adjudicator.

Parties must comply with the decision of the Adjudicator; and the Employer and the Contractor shall ensure that the decision is given effect. If either Party does not comply with the decision of the Adjudicator, the other Party is entitled to take legal proceedings to secure such compliance.

41A.8 The Adjudicator is given immunity from liability for anything done, or omitted, in the discharge or purported discharge of his functions as Adjudicator unless the act or omission is in bad faith. This is an important immunity which will allow Adjudicators to make decisions with confidence.

Clause 41B: Arbitration

41B.1 Any reference in this clause to a rule or rules is a reference to the JCT 1998 edition of the Construction Industry Model Arbitration Rules (CIMAR) current at the base date.

Where either Party requires a dispute or difference to be referred to Arbitration then that Party shall serve on the other party a notice of Arbitration in accordance with Rule 2.1:

Arbitral proceedings are begun in respect of a dispute when one Party serves on the other Party a written notice of Arbitration identifying the dispute and requiring him to agree to the appointment of an Arbitrator.

An Arbitrator shall be an individual agreed by the parties or appointed by the person named in the Appendix in accordance with Rule 2.3 which states:

If the parties fail to agree on a name of an Arbitrator within 14 days (or any agreed extension) after:
the notice of arbitration is served, or
a previously appointed Arbitrator ceases to hold office for any reason
either party may apply for the appointment of an Arbitrator to the person so empowered.

The normal procedure is that the parties agree between themselves to the Arbitrator, or when they cannot agree the President of the Royal Institute of British Architects or the Royal Institution of Chartered Surveyors or the Chartered Institute of Arbitrators appoints one.

Where two or more related arbitral proceedings fall under separate agreements they may be heard together (Rules 2.6, 2.7 and 2.8). This is an attempt to make Arbitration, like litigation, appropriate for multi-party issues by joinder and to avoid the need for many similar but different proceedings.

41B.2 The Arbitrator is given the wide power to rectify the contract so as to reflect the true agreement between the parties.

41B.3 This clause reinforces the final and binding nature of the Arbitrator's Award. An Arbitrator's Award can be subject to appeal on limited grounds via case law and the Arbitration Act 1996.

41B.4 The parties agree pursuant to the Arbitration Act 1996 to:
apply to the Court for the determination of any question of law arising in the Arbitration;
apply to the Court for the determination of any question of law arising out of the Arbitration Award.

41B.5 The provisions of the Arbitration Act 1996 apply.

41B.6 The Arbitration shall be conducted in accordance with the Construction Industry Model Arbitration Rules (CIMAR).

Clause 41C: Legal Proceedings

The traditional provision for dispute resolution in JCT contracts has been arbitration. JCT 98 includes provision for any dispute or difference to be determined by legal proceedings and the decision between arbitration or legal proceedings is made in the Appendix. There are many reasons for the alternatives; there was a feeling that historically arbitration clauses were often deleted from contracts and that an option should exist. For a long period Arbitrators had wider powers than Judges in court following the decision in *Northern Regional Health Authority* v. *Derek Crouch Construction Company Limited* (1984). Thus implications of Crouch have been removed with the decision in *Beaufort Developments Limited* v. *Gilbert Ash NI Limited* (1998) which overturned the decision in Crouch, which now means that Judges and Arbitrators have the same powers.

Clauses 41A–41C: Further explanation

JCT 98 refers to four methods of dispute resolution:

Mediation
Adjudication
Arbitration
Legal proceedings

As described, Adjudication, Arbitration, and legal proceedings are dealt with in Part 4: Clause 41. Mediation is only referred to in a footnote: 'It is open to the Employer and the Contractor to resolve disputes by the process of Mediation: see Practice Note 28 "Mediation on a Building Contract or Sub-contract Dispute".' This further explanation deals with the techniques initially in terms of a distinction between conflict and dispute.

Conflict and Dispute

This section is concerned with the techniques used to resolve disputes when the parties to a JCT contract have been unable to settle a difference. Distinctions might be made between the steps to avoid conflict and dispute, the parties own attempts to manage conflict and the techniques of dispute resolution. This distinction is shown diagrammatically below where the conflict and dispute split is shown as a continuum.

The continuum shows a distinct gap between conflict avoidance and informal discussion; conflict avoidance might be the steps taken before a Contract is drawn up. Then a further distinction is made between negotiation and Alternative Dispute Resolution (ADR), often the intervention of a third party. The dispute resolution phase runs from ADR to litigation. There is a phase beyond and hopefully we will not have to confront the danger of violence!

Often the distinction between conflict and dispute involves the intervention of a third party; the techniques of dispute resolution involve the intervention of: a conciliator or mediator; an Adjudicator, an arbitrator or a judge.

Mediation

Mediators are reluctant to provide decisions or recommendations, indeed some mediators will not do so. Mediation lies at the facilitative end of the spectrum. Any settlement that occurs is the parties' own and the analogy of chemical catalysts is often made. A catalyst makes a reaction take place between two or more chemicals; the catalyst is not affected or changed by the reaction. Sometimes the reaction will take place without the

Resolution → Dispute

Conflict Management →

Conflict Avoidance	Informal Discussion	Negotiate	ADR	Arbitrate Adjudicate	Litigate	Other Action Violence

DISPUTE

CONFLICT

Construction Conflict Continuum

catalyst and the effect is to speed it up, and sometimes the reaction will not take place without the catalyst. The analogies with mediation are obvious: a mediator may facilitate or speed things up, or the mediator may bring about a settlement which otherwise may not have happened.

Mediation is the most widely used and accepted ADR technique. While there is no prescriptive mediation process, the typical stages in a mediation might be:

- A brief written summary of the matter in dispute is presented in advance to the mediator.
- The parties meet with a mediator for an initial joint meeting, including perhaps a brief oral presentation by the parties.
- Caucus sessions, where the mediator has private meetings with the party in turn. During the caucuses, the mediator often shuttles backwards and forwards to clarify issues and search for settlement possibilities. This process is often termed shuttle diplomacy.
- Plenary sessions are called either to continue negotiations directly, to conclude agreement, or where the process is unsuccessful to conclude a mediation.

Most mediators agree to a contingency approach to mediation; that is, there is no set procedure but the procedure is tailored to suit the parties and the dispute in question. This often means that mediation is conducted without joint meetings and the mediators play a variety of roles. The mediator may act as a mere facilitator, there purely to assist communications. Alternatively the mediator acts as a deal maker, to assist the parties in finding overlap in their bargaining positions or encouraging concession and compromise. Perhaps the mediator acts more as a problem-solver, assisting the parties in designing and searching for creative solutions. The mediator may act as a transformer, transforming the dispute by allowing the parties a fresh insight into the issues and their positions. The final role of the mediator may be as an Adjudicator or assessor to provide the parties with an appraisal of the merits of the cases on a legal, technical or even common-sense standpoint.

Joint Contracts Tribunal (JCT) and Mediation

The JCT considered the application of ADR to its forms of contract for construction as the ADR movement strengthened. The first moves were made by the National Joint Consultative Committee (NJCC) in *NJCC Guidance Note 7, Alternative Dispute Resolution,* January 1993; and in 1995 the JCT issued *JCT Standard Forms of Building Contract Practice Note 28, Mediation on a Building Contract or Sub-Contract Dispute, PN 28/95.* PN 28 reminded the parties that a variety of dispute resolution techniques is available and confirmed the status of Mediation as a facilitated negotiation. PN 28 included the following drafts for use on either building contracts or sub-contracts:

Mediation Agreement
Agreement appointing a Mediator
Agreement following the resolution of a dispute after a Mediation

In addition, the following text was included as a footnote to the provisions setting out the arbitration agreement:

> *It is open to the Employer and the Contractor to the Contractor and the Sub-Contractor to resolve disputes by the process of Mediation: see Practice Note 28, Mediation on a Building Contract or Sub-Contract Dispute.*

PN 28 provides little in the way of detail of any mediation but suggests that the procedure will be a facilitative one carried out by an independent person, whose qualifications will depend on the dispute, appointed not to impose a solution but to try to steer the parties themselves towards a settlement. It is suggested that whilst Mediation should always be considered it may not be appropriate where the following factors are present:

- The dispute is one where a legal precedent is being sought.
- Either party needs an injunction to preserve the *status quo* pending a legal decision on the dispute.
- One of the parties wants a public hearing.

- One or other party is not genuine in wanting to reach a settlement by agreement.

The guidance on mediation agreements states that the main matters to be considered are:

- A clear and precise statement of the issues in the dispute.
- A declaration of the parties' wishes to resolve the dispute with the help of the mediator.
- The period during which the Mediation is to take place but leaving either party free to withdraw without giving reasons and so bringing the Mediation to an end; or for the parties to extend the period.
- The name and qualifications of the Mediator.
- Provisions that the Mediation is to be conducted on a confidential and without prejudice basis.
- A date for meeting the Mediator to discuss among other things the methods by which the Mediation is to be conducted.
- The question of how costs will be dealt with (the guidance is that the parties bear their own costs but share the fee of the Mediator equally).
- Provision that if the Mediation results in a settlement the parties execute a binding agreement setting out the terms of the settlement.

PN 28 reminds the parties that there may be a requirement to proceed with the Contract as if there were no dispute and that an arbitration under Rule 7 of the JCT Arbitration Rules (short procedure without a hearing) might more readily effect this.

Adjudication

Prior to 1976 no standard forms of construction contract contained Adjudication provisions. The traditional method of resolving disputes was either litigation or arbitration, if a binding arbitration agreement existed. Initially Adjudication was introduced to deal with set-off.

Set-Off/Adjudication Clauses were introduced as follows:

1976 JCT 63 Sub-contract forms (Green and Blue Forms).
1980 NSC4/4A Forms for use with JCT 80 Standard (DOM1) Form of Main contract.
1981 DOM2 Sub-contract for use with JCT 81 with Contractors Design Main contract.
1984 NAM/SC and IN/SC named and domestic Sub-contract forms for use with Intermediate form of contract (IFC 84).
1987 JCT Works contract for use with JCT Management contract.

These were then extended in their application to other issues and other contract forms were made available:

1983 BPF form of Building contract – Adjudication provided in respect of a wide range of issues beyond set-off claims. Optional Adjudication was available in its precursor form, The ACA form of Building contract 1982.

1985 GW/S Sub-contract for use with GCWorks 1, Edition 2, 1985 (set-off/Adjudication clause similar to NSC4/DOM1 Sub-contract).

1988 JCT 81 With Contractors Design Main contract – Adjudication is available in relation to a wide range of issues.

1989 GC Works 1, Edition 3 (not limited to set-off).

1991 ICE Draft New Engineering contract, Adjudication of all disputes as an interim procedure.

The concept of contractual Adjudication for construction projects was championed by The Latham Report (1994) and the New Engineering and Construction contract included a wide-ranging Adjudication clause. The government confirmed its commitment to the idea by including the Latham proposals in legislation by means of *The Housing Grants, Construction and Regeneration Act 1996 (HGCR) via the Scheme for Construction contracts*. Here the legislation provides that all construction contracts must include a provision for Adjudication of disputes within defined timescales; if the contract does not provide such provisions then the default provisions of the Scheme will apply.

Adjudication as dispute resolution presents a problem for the construction industry. The problem lies in Adjudication as an established practice, and Adjudication as envisaged by The Latham Report and provided for by HGCR and The Scheme for Construction contracts.

Adjudication has existed as established construction practice for some time. Standard forms have included express provision for Adjudication, although the grounds for such Adjudication were normally restricted to payment and *set-off* matters.

Early debate on the subject of Adjudication merely treated Adjudication from the position of what it is **not**. Adjudication is not arbitration. Adjudication was classified together with the 'other' Adjudicative binding processes, such as reference to an expert, independent valuation and expert determination.

The HGCR Act gives little guidance on the intended nature of the Adjudication process, although benchmarks for timescales are laid down in The Scheme for Construction contracts. Adjudication is clearly distinguished from arbitration, for example:

'An exchange of written submissions in Adjudication proceedings **or** in arbitral or legal proceedings' [Section 107(3)]

'The contract shall provide that the decision of the Adjudicator is binding until the dispute is finally determined by . . . arbitration'

[Section 108(3)]

Adjudication Provisions of the HGCR

Under HGCR it is now mandatory for a party to have the right to refer a dispute under a construction contract to Adjudication at any time. This right cannot be excluded or contracted out of. Where the Act provides that the parties can agree on a particular matter covered by the Act, it will provide for certain minimum requirements for the agreement to be valid and/or will refer to the Scheme if the parties have not agreed, or if their agreement does not meet minimum requirements. The Act itself sets out the skeleton of the Adjudication procedure and the requirements of every construction contract.

The parties to a contract can have their own scheme for Adjudication in respect of a dispute arising under the contract, which is valid provided it meets certain minimum requirements. Section 108 details as follows these minimum requirements that a contract must provide for:

- Notice at any time of intention to refer a dispute to Adjudication.
- A timetable with the object of securing the appointment of the Adjudicator and referral of the dispute to the Adjudicator within seven days of the notice.
- The Adjudicator's decision to be within 28 days of referral, although the period can be extended by agreement of the parties after the referral.
- The Adjudicator is allowed to extend the 28-day period by up to 14 days with the consent of the referring party.
- The Adjudicator is required to act impartially.
- The Adjudicator is allowed to take the initiative in ascertaining the facts and the law.
- The Adjudicator's decision to be binding until the dispute is finally determined by legal action, by arbitration (if provided for/agreed to), or by agreement, although the parties may agree to accept the decision as finally determining the dispute.
- The Adjudicator has immunity from anything done or omitted while acting as Adjudicator, unless done in bad faith.

If a contract does not comply with any one of these requirements, Part 1 of the Scheme, relating to Adjudication, will apply in total, overriding even those parts of the contract which do comply.

In summary it is worth realising that while referral of a dispute to Adjudication is optional rather than mandatory, all construction contracts caught by the Act must provide for Adjudication.

The Adjudication provisions within the Act are brief and silent on many matters regarding the Adjudication process. The Scheme fills some of these gaps, as do the sets of Adjudication rules issued by various construction industry bodies.

Adjudication and the Courts

The role of Adjudication under the Act and Scheme has been dealt with by case law and the courts have by many decisions dealt with Adjudicator's awards and Adjudication procedure.

Cape Durasteel Ltd v. *Rosser and Russell Building Services Ltd* (1996) considered the use of the word Adjudication in the dispute procedure provided for in the contract:

> *'It is plain that "Adjudication" taken by itself means a process by which a dispute is resolved in a judicial manner. It is equally clear that "Adjudication" has yet no settled special meaning in the construction industry (which is not surprising since it is a creature of contract and contractual procedures utilising an "Adjudicator" vary as do forms of contract). Even if it were to have the special meaning accorded to it in some sections of the construction industry where it describes the initial determination of certain classes of dispute in a summary manner, the force of which is tempered by its ephemeral status as there are concomitant provisions for the decision to be reviewed and if necessary reversed by an arbitrator, I would see no reason why it should have that special meaning in this contract.'*

Macob Civil Engineering Ltd v. *Morrison Construction Ltd* (1999):

> *'The intention of Parliament in enacting the Act was plain. It was to introduce a speedy mechanism for settling disputes in construction contracts on a provisional interim basis . . .'*

Macob demonstrated the courts' intention to support Adjudication. This brought Adjudication to the forefront of the dispute resolution process with the message that the courts would be likely to treat an Adjudicator's decision as binding, and would enforce it summarily pending final resolution of the dispute.

Outwing Construction Ltd v. *H Randell and Sons Ltd* (1999) reinforced the view taken in Macob.

A&D Maintenance and Construction Ltd v. *Pagehurst Construction Services Ltd* (1999) provided further confirmation of the courts' support for statutory Adjudication. The court granted enforcement of an Adjudicator's decision by way of an application for summary judgment.

The Project Consultancy Group v. *The Trustees of the Gray Trust* (1999) again involved the enforcement of an Adjudicator's decision by summary judgment, but on this occasion was unsuccessful.

A Straume (UK) Ltd v. *Bradlor Developments Ltd* (2000) dealt with whether or not Adjudication proceedings, because Bradlor had been placed in administration, are 'other proceedings' within the meaning of Section 11(3) of the Insolvency Act 1986 and therefore subject to leave of the court being required to commence proceedings. It was decided that they are and in this particular case leave was not granted to commence an Adjudication.

John Cothliff v. *Allen Build (North West) Ltd* (1999). Cothliff sought payment of their costs of the Adjudication. The Adjudicator, after considering that the provisions of the Scheme for Construction Contracts gave the power to do this, awarded Cothliff a proportion of their costs. Allen Build refused to comply with the Adjudicator's decision in regard to the costs' element and Cothliff commenced proceedings for summary judgment. This was subsequently overturned in the case of *Northern Developments (Cumbria) Ltd* v. *J&J Nichols* (2000).

Bouygues UK Ltd v. *Dahl-Jensen UK Ltd* (2000) involved the enforcement of an Adjudicator's decision by summary judgment and further demonstrated the courts support of such enforcement even when the Adjudicator's decision was clearly wrong.

Bloor Construction (UK) Ltd v. *Bowmer & Kirkland (London) Ltd* (2000) involved an Adjudication in which the Sub-Contractor, Bloor, was initially awarded a sum that had overlooked the on-account payments made to-date by the main Contractor; the court took into account the error made by the Adjudicator. It was decided that in the absence of any specific agreement to the contrary, a term can be implied into the underlying contract between the parties referring the dispute to Adjudication: that the Adjudicator may, on his own initiative or on the application of a party, correct an error arising from an accidental error or omission. The judge said that the *'purpose of the Adjudication is to enable broad justice to be done between the parties.'* The court appears to have found a clever way round the fact that the court cannot directly correct obvious errors in Adjudications. Perhaps the courts can also say that manifest error should be capable of being effectively challenged in the courts by this contractual route.

Herschel Engineering Ltd v. *Breen Property Ltd* (2000) dealt with the question of whether or not Adjudication could be pursued at the same time as court proceedings in respect of the same subject matter. The Judge explained that Parliament's intention in enacting the Act was to produce a speedy mechanism for resolving disputes on a provisional interim basis and that Adjudicators' decisions were to be enforced pending final determination. If Parliament had intended that a party should not be able to refer a dispute to Adjudication once litigation or arbitration had commenced it would have been expressly stated.

Bridgeway Construction Ltd v. *Tolent Construction Ltd* (11 April 2000), unreported. A main Contractor incorporated into the sub-contract various special conditions relating to Adjudication. These conditions were that the applicant party on an Adjudication should pay in effect not only its own costs of the Adjudication but also the Adjudicator's fees and also the costs of the defending party. The Sub-Contractor in question commenced an Adjudication and the Adjudicator awarded the Sub-Contractor certain monies but also ordered that the Sub-Contractor pay its own costs, the Adjudicator's fees and the defending main Contractor's costs. When the Sub-Contractor came to seek to enforce the decision in the courts, the main Contractor maintained it had paid the net sums outstanding under the Adjudication, having deducted the amount for the defendant's costs. It was held that this was a wholly legitimate way for the parties to have legislated that contractual relationship, as a matter of contract, and it now appears that the parties can contractually secure that Adjudication is an unattractive route for Sub-Contractors.

F W Cook v. *Shimizu* (2000 BLR 199) is an example of the importance that an analysis of the referral notice can be. Cook were Sub-Contractors engaged by Shimizu. Cook gave notice of intention to seek Adjudication in the context of a dispute about a final account. Having identified what its provisional final account was, Cook sought to refer four out of a number of final account disputes to an Adjudicator. The Adjudicator indicated in his decision that on three of the four matters Cook was entitled to be paid or to recover certain sums. Shimizu only paid some 10% of the amount. Shimizu had adjusted other matters on the final account to produce that lower figure. The Court considered carefully the referral notice and the text of the Adjudicator's decision:

> *'The referring letter expresses the view on the part of Cook that, out of the many items in dispute on the final account, it wished to have decisions on certain of them in the hope that other items might well be resolved once the Adjudicator's decisions were given. [. . .] Looking at this letter and notice it is plain that Cook wished to obtain a decision on what might not correctly be called a principle, but which were a number of items or elements or ingredients in an overall final account and valuation, as opposed to obtaining a decision as to how much the next interim payment should be.'*

The judge effectively found that the Adjudicator had had regard to the referral notice properly and had in effect decided that the sums identified as payable in the decision were not necessarily immediately payable. If the Adjudicator does more than the referral notice permits him to do he will be acting in excess of jurisdiction.

Arbitration

In England and Wales, arbitration is governed by legislation. The first legislation was the Arbitration Act 1698, culminating in the Arbitration Act 1996. The parties entering into Arbitration are bound to accept the award as final and binding upon them; and the award is enforceable as a judgement of the court.

The Arbitration Act 1996 specifically avoids any definition but commences with a statement of general principles:

1. *The object of arbitration is to obtain the fair resolution of disputes by an impartial tribunal without unnecessary delay or expense;*
2. *The parties should be free to agree how their disputes are resolved, subject only to such safeguards as are necessary in the public interest;*
3. *. . . the courts should not intervene . . .*

Arbitration has been traced back to ancient Greece, although most sources credit the Romans with originating arbitration via a Praetor. The praetorial system of arbitrary justice was carried throughout the Roman Empire and found favour with merchants and traders in England. This law of the merchants was eventually administered via the Courts of Pipowder; a corruption of the French *Pieds Poudres* – the court of the dusty feet.

The Industrial Revolution and the advent of limited liability brought about the formation of corporate bodies. Once again the civil courts failed to serve the developments. Following the Industrial Revolution it became apparent that the procedures of the civil courts were hardly the ideal place for the resolution of corporate disputes. Business complained of the vagaries of the system, and took their disputes elsewhere – notably the arbitrators' room. The construction industry in particular was quick to take on board the developing process of arbitration. The formation, in 1915, of the Chartered Institute of Arbitrators included several construction professionals on the founding committee.

Arbitration developed to be the preferred method of resolving disputes in the construction industry and in other areas, notably shipping. Litigation and arbitration continued in an atmosphere of mutual distrust; occasionally legislation enforces arbitration as the method of resolving disputes. All the widely adopted standard forms of contract for construction works include an arbitration agreement; and arbitration is generally seen as the common method for resolving disputes.

The major advantages proposed for arbitration are:

- Speed of proceedings
- Cost of proceedings
- Privacy
- Technical expertise of arbitrator.

Arbitration has been accused of suffering from the same problems which afflicted the civil courts after the Industrial Revolution: the very problems which provided the environment for arbitration to flourish. The perceived advantages of low cost, high speed, and availability of technical expertise have all but disappeared as arbitrators have increasingly mimicked the High Court for arbitration procedures. Parties are very frequently represented by senior counsel, the rules for evidence and proof follow High Court procedure, and proceedings have become protracted. Parties to arbitration increasingly turned to the courts for appeals against the decision of their arbitrators.

The Arbitration Act 1996 has attempted to deal with these problems and initial reports confirm that the new powers given to Arbitrators are being used. It may be that Adjudication via HGCR will reduce the number of Arbitrations.

Arbitration or Litigation?

The choice between litigation or arbitration was influenced by precedent which gave arbitrators powers not available to judges, unless the parties specifically gave judges such powers. These powers were described as Crouch Powers following the decision in *Northern Regional Health Authority* v. *Derek Crouch Construction Company Limited* (1984). This case was felt to have limited the jurisdiction of judges, or at least to have defined it. The result was that parties to a contractual dispute were more likely to refer the matter to an Arbitrator, since the Arbitrator alone has the power to open up and review certificates. The Crouch case arose out of contracts between:

- Employer/Building owner: The Health Authority
- Main Contractor: Derek Crouch
- Sub-Contractor: Crown House Engineering

The contract was the Standard Form of Building Contract issued by the Joint Contracts Tribunal, 1963 edition (JCT 63), which contains an arbitration clause giving the Arbitrator power: 'to open up, review and revise any certificate, opinion, decision, requirement or notice and to determine all matters in dispute which shall be submitted to him in the same manner as if no such certificate, opinion, decision, requirement or notice had been given.'

Disputes arose between the parties; some were the subject of arbitration and some litigation. The decision on the case was by the Court of Appeal. The Health Authority attempted, by originating summonses to the judge, to secure an injunction restraining Crouch and Crown from seeking awards in arbitration. It was held that this would amount to an injustice for a variety of reasons, including the fact that the judge would not have the powers of an Arbitrator in accordance with the arbitration clause. This was contrary to the position adopted by the judges, who had in their frequent dealings with JCT, and indeed all construction contracts, assumed their powers were equal to an Arbitrator.

This implications of Crouch have been removed with the decision in *Beaufort Developments Limited* v. *Gilbert Ash NI Limited* [1998] which overturned the decision in Crouch. The result may be a move away from Arbitration of Litigation.

Conditions: Part 5: Performance Specified Work

Clause 42: Performance Specified Work

See also: Practice Note 25

42.1 'Performance Specified Work' is work:

    .1   identified in the Appendix, and

    .2   to be provided by the Contractor, and

    .3   for which predetermined requirements are shown on the Contract Drawings, and

    .4   the Contract Bills state the performance required either:

        (a) by including sufficient information to allow the Contractor to price the work, or

        (b) by including a provisional sum for the work in accordance with Clause 42.7.

    .2   The Contractor must provide the Architect with a 'Contractor's Statement', in accordance with which the Performance Specified Work will be executed, prior to carrying out any such work.

    .3   The Contractor's Statement must explain how the Contractor proposes to execute the Performance Specified Work, include information required by the BQ and may include drawings, schedules and calculations.

    .4   The Contractor's Statement must be provided to the Architect at a reasonable time before the intended commencement of the Performance Specified Work unless a date for its provision is given in:

        (a) The BQ, or

        (b) The AI re the provisional sum.

.5   If the Architect believes that the Contractor's Statement is deficient, he has 14 days to give written notice to the Contractor requiring amendment of the statement.

.6   If the Architect finds any deficiency in the Contractor's Statement which he believes would affect the performance of the Performance Specified Work adversely, he must give notice specifying the deficiency to the Contractor immediately.

Despite any notice under clauses 42.5 or 42.6, the Contractor retains full responsibility for the Statement and for the Performance Specified Work.

.7   A provisional sum in the BQ for Performance Specified Work must provide:

.1   the performance which the Employer requires,

.2   the location of the Performance Specified Work,

.3   sufficient information for the Contractor to make appropriate allowance in programming the work and for pricing preliminaries.

.8   No AI can require Performance Specified Work except one in respect of expenditure of a provisional sum included in the BQ for Performance Specified Work.

.9   Inclusion of Performance Specified Work is not to be regarded as a departure from the method of preparing the Contract Bills under Clause 2.2.2.1.

.10   Errors/omissions in the BQ regarding Performance Specified Work – to be corrected and such correction treated as a Variation under Clause 13.2.

.11   AIs of Variation (Clause 13.2) may be issued in respect of Performance Specified Work, subject to Clause 42.12.

.12   Unless the Employer and Contractor so agree, no AI may require Performance Specified Work additional to that identified in the Appendix.

.13   Unless the Contract Bills provide an analysis of the portion of the Contract Sum which relates to Performance Specified Work, the Contractor must provide such an analysis (the 'Analysis') within 14 days of being required to do so by the Architect.

.14   The Architect must give any AIs for the integration of Performance Specified Work with the design of the Works within a reasonable time before the Contractor intends to execute such work. Subject to Clause 4.2.15, the Contractor must comply with such AIs.

.15   If the Contractor believes that compliance with any AI affects the efficacy of Performance Specified Work injuriously, he must give written notice of such affection to the Architect within 7 days of receipt of the AI.

The AI shall have no effect without the Contractor's written consent (not unreasonably withheld) unless the Architect amends the AI to remove the injurious affection.

.16   If the Architect does not receive the Contractor's Statement in time under Clause 42.4, or any amendment(s) thereto under Clause 42.5, no extension of time (Clause 25.3 or loss/expense (Clauses 26.1 and 28.2.2) may be awarded for consequent delays/expenses, etc. caused thereby. The exception is in respect of Relevant Events under Clause 25.4.15 (Statutory Requirements after the Base Date requiring changes to Performance Specified Work).

.17   .1   The Contractor must exercise reasonable skill and care in the provision of Performance Specified Work.

.1 Clause 42.17 does not affect the overriding obligations of the Contractor regarding workmanship, etc.

.2 Nothing in the Contract operates as a guarantee of fitness for purpose of Performance Specified Work.

.2 The Contractor retains full responsibility for any sub-contracting, etc. in respect of any aspect of Performance Specified Work.

.18   Performance Specified Work (under Clause 42) must not be provided by a NS/C or a NSup.

Code of Practice: Referred to in Clause 8.4.4

1   The Code of Practice is referred to in Clause 8.4.4; its purpose is to assist in the fair and reasonable operation of the requirements of that clause (extent of inspection/testing where some items do not accord with the Contract).

2   The Architect and Contractor should try to agree the extent and method of opening up and testing of work.

In any case, the Architect must consider the criteria noted in the Code of Practice in issuing AIs under Clause 8.4.4:

.1   In the event of non-compliance (of work, etc. with the contract) the need to demonstrate that such non-compliance is either:

(a) unique and not likely to occur in similar elements of the Works, or

(b) the extent of similar non-compliance in Works constructed already or still to be constructed.

Such demonstration must be at no cost to the Employer.

.2   'The need to discover whether any non-compliance in a primary structural element is a failure of workmanship and/or materials such that similar testing of similar elements must take place.'

If non-compliance is in a less significant element, whether it is expected statistically and can be repaired easily or whether the non-compliance indicated an inherent weakness/problem, the extent of which can be revealed by selective testing; the scope of action required must depend upon the importance of the element and likely consequences of the non-compliance.

Note: This clause implies that while a statistical approach to testing is appropriate for non-primary structured elements; if a failure is found in primary structural elements, a full population test is necessary – i.e. test all such elements!

Given the major principles established in .1 and .2, the following must be considered also:

188

.3 the significance of the non-compliance, considering the work in which it was found,

.4 consequences of further instances of non-compliance on:

(a) safety of the building,

(b) effects on users,

(c) adjoining property,

(d) the public,

(e) Compliance with Statutory Requirements.

.5 Supervision and Control of the Works by the Contractor (note changes therein relevant to the incidence of non-compliance and probability of further such occurrences),

.6 records of Contractor and Sub-Contractors,

.7 Codes of Practice, etc. relating to the non-complying work (i.e. were such codes etc. followed?),

.8 any failure(s) by the Contractor to effect the tests specified in the Contract Documents,

.9 reason(s) for the non-compliance,

.10 any technical advice obtained by the Contractor regarding the non-compliance,

.11 'current recognised testing procedures',

.12 'the practicability of progressive testing in establishing whether any similar non-compliance is reasonably likely,

.13 time for and costs of alternative methods of testing, if available,

.14 any Contractor's proposals,

.15 other relevant matters.

So, apart from primary structural elements, statistical approaches and consideration of advice and alternative tests must be employed.

Appendix

The provisions of the Appendix are quite self-explanatory and have been referred to throughout the consideration of the Conditions.

The matters to be dealt with in the Appendix are listed below.

Clause etc.	Subject	
Fourth recital and 31	Statutory tax deduction scheme	Employer at Base Data *is a 'contractor'/is not a 'contractor' for the purposes of the Act and the Regulations
Fifth recital	CDM Regulations	*All the CDM Regulations apply/ Regulations 7 and 13 only of the CDM Regulations apply
Articles 7A and 7B 41B 41C	Dispute or difference – settlement of disputes	*Clause 41B applies
		*Delete if disputes are to be decided by legal proceedings and article 7B is thus to apply
1.3	Base Date	_____
1.3	Date for Completion	_____
1.11	Electronic data interchange	The JCT Supplemental Provisions for EDI *apply/do not apply
		If applicable: the EDI Agreement to which the Supplemental Provisions refer is: *the EDI Association Standard EDI Agreement *the European Model EDI Agreement

Footnotes   *Delete as applicable.

Clause etc.	Subject	
15.2	VAT Agreement	Clause 1A of the VAT Agreement *applies/does not apply [1]
17.2	Defects Liability Period (if none other stated is 6 months from the day named in the certificate of Practical Completion of the Works)	_____
19.1.2	Assignment by Employer of benefits after Practical Completion	Clause 19.1.2 *applies/does not apply
21.1.1	Insurance cover for any one occurrence or series of occurrences arising out of one event	£ _____
21.2.1	Insurance – liability of Employer	Insurance *may be required/is not required
		Amount of indemnity for any one occurrence or series of occurrences arising out of one event
		£ _____ [2]

Clause etc.	Subject	
22.1	Insurance of the Works – alternative clauses	*Clause 22A/Clause 22B/Clause 22C applies (See footnote [cc] to clause 22)
*22A, 22B.1, 22C.2	Percentage to cover professional fees	_____
22A.3.1	Annual renewal date of insurance as supplied by Contractor	_____
22D	Insurance for Employer's loss of liquidated damages – clause 25.4.3	Insurance *may be required/is not required
22D.2		Period of time _____
22FC.1	Joint Fire Code	The Joint Fire Code *applies/does not apply If the Joint Fire Code is applicable, state whether the insurer under clause 22A or clause 22B or clause 22C.2 has specified that the Works are a 'Large Project': *YES/NO (where clause 22A applies these entries are made on information supplied by the Contractor)
23.1.1	Date of Possession	_____
23.1.2, 25.4.13, 26.1	Deferment of the Date of Possession	Clause 23.1.2 *applies/does not apply Period of deferment if it is to be less than 6 weeks is _____
24.2	Liquidated and ascertained damages	at the rate of £ _____ per _____

Footnotes   *Delete as applicable.

Clause etc.	Subject	
28.2.2	Period of suspension (if none stated is 1 month)	_____
28A.1.1.1 to 28A.1.1.3	Period of suspension (if none stated is 3 months)	_____
28A.1.1.4 to 28A.1.1.6	Period of suspension (if none stated is 1 month)	_____

30.1.1.6   Advance payment

Clause 30.1.1.6
*applies/does not apply

If applicable:
the advance payment will be

**£ ————————————— /

——————— % of the Contract Sum and

will be paid to the Contractor on

_____

and will be reimbursed to the
Employer in the following
amount(s) and at the following
time(s)

_____

_____

_____

An advance payment bond
*is/is not required

| 30.1.3 | Period of Interim Certificates (if none stated, Interim Certificates are to be at intervals not exceeding one month) | _____ |

Footnotes   *Delete as applicable.

Clause etc.	Subject	
30.2.1.1	Gross valuation	A priced Activity Schedule *is/is not attached to this Appendix
30.3.1	Listed items – uniquely identified	*For uniquely identified listed items a bond as referred to in clause 30.3.1 in respect of payment for such items is required for £ _____ *Delete if no bond is required
30.3.2	Listed items – not uniquely identified	*For listed items that are not uniquely identified a bond as referred to in clause 30.3.2 in respect of payment for such items is required for £ _____ *Delete if clause 30.3.2 does not apply
30.4.1.1	Retention Percentage (if less than 5 per cent) [3]	_____
35.2	Work reserved for Nominated Sub-Contractors for which the Contractor desires to tender	_____
37	Fluctuations: (if alternative required is not shown clause 38 shall apply)	clause 38 [4] clause 39 clause 40
38.7 or 39.8	Percentage addition	_____

Footnotes

*Delete as applicable.

**Insert either a money amount or a percentage figure and delete the other alternative.

[3] The percentage will be 5 per cent unless a lower rate is specified here.

[4] Delete alternatives not used.

Clause etc.	*Subject*	
40.1.1.1	Formula Rules	rule 3: Base Month
		_____ 19 _____
		rules 10 and 30 (i): Part I/Part II [5] of Section 2 of the Formula Rules is to apply
41A.2	Adjudication – nominator of Adjudicator (if no nominator is selected the nominator shall be the President or a Vice-President of the Royal Institute of British Architects)	President or a Vice-President or Chairman or a Vice-Chairman: *Royal Institute of British Architects *Royal Institution of Chartered Surveyors *Construction Confederation *National Specialist Contractors Council

*Delete all but one |
| 41B.1 | Arbitration – appointor of Arbitrator (if no appointor is selected the appointor shall be the President or a Vice-President of the Royal Institute of British Architects) | President or a Vice-President: *Royal Institute of British Architects *Royal Institution of Chartered Surveyors *Chartered Institute of Arbitrators

*Delete all but one |
| 42.1.1 | Performance Specified Work | Identify below or on a separate sheet each item of Performance Specified Work to be provided by the Contractor and insert the relevant reference in the Contract Bills [6] |

Footnotes

[5] Strike out according to which method of formula adjustment (Part I – Work Category Method or Part II – Work Group Method) has been stated in the documents issued to tenderers.

[6] See Practice Note 25 'Performance Specified Work' paragraphs 2.6 to 2.8 for a description of work which is **not** to be treated as Performance Specified Work.

Annex 1 to Appendix: Terms of Bond

The text of a bond agreed between the British Bankers' Association and the JCT is shown below.

Appropriate clause 30.1.1.6:
'Advance Payment Bond', and

Appropriate clause 30.3:
'Bond in respect of payment for off-site materials and/or goods'

Advance Payment Bond

1   THE parties to this Bond are:

(1) _____

whose registered office is at _____

_____ ('the Surety'), and

(2) _____

of _____

_____ ('the Employer').

2   The Employer and _____ ('the Contractor')

have agreed to enter into a contract for building works ('the

Works') at _____

_____ ('the Contract').

3   The Employer has agreed to pay the Contractor the sum of [_____]
     as an advance payment of sums due to the Contractor under the

196

Contract ('the Advance Payment') for reimbursement by the Surety on the following terms:

(a) When the Surety receives a demand from the Employer in accordance with clause 3(b) the Surety shall repay the Employer the sum demanded up to the amount of the Advance Payment.

(b) The Employer shall in making any demand provide to the Surety a completed notice of demand in the form of the **Schedule** attached hereto which shall be accepted as conclusive evidence for all purposes under this Bond. The signatures on any such demand must be authenticated by the Employer's bankers.

(c) The Surety shall within 5 Business Days after receiving the demand pay to the Employer the sum so demanded. 'Business Day' means the day (other than a Saturday or a Sunday) on which commercial banks are open for business in London.

4   Payments due under this Bond shall be made notwithstanding any dispute between the Employer and the Contractor and whether or not the Employer and the Contractor are or might be under any liability one to the other. Payment by the Surety under this Bond shall be deemed a valid payment for all purposes of this Bond and shall discharge the Surety from liability to the extent of such payment.

5   The Surety consents and agrees that the following actions by the Employer may be made and done without notice to or consent of the Surety and without in any way affecting changing or releasing the Surety from its obligations under this Bond and the liability of the Surety hereunder shall not in any way be affected hereby. The actions are:

(a)   waiver by the Employer of any of the terms, provisions, conditions, obligations and agreements of the Contractor or any failure to make demand upon or take action against the Contractor;

(b)   any modification or changes to the Contract; and/or

(c)   the granting of any extensions of time to the Contractor without affecting the terms of clause 7(c) below.

6   The Surety's maximum aggregate liability under this Bond which shall commence on payment of the Advance Payment by the

Employer to the Contractor shall be the amount of [_____] which sum shall be reduced by the amount of any reimbursement made by the Contractor to the Employer as advised by the Employer in writing to the Surety.

7   The obligations of the Surety and under this Bond shall cease upon whichever is the earliest of:

(a) the date on which the Advance Payment is reduced to nil as certified in writing to the Surety by the Employer;

(b) the date on which the Advance Payment or any balance thereof is repaid to the Employer by the Contractor (as certified in writing to the Surety by the Employer) or by the Surety; and

(c) *[longstop date to be given]*,

and any claims hereunder must be received by the Surety in writing on or before such earliest date.

8   This Bond is not transferable or assignable without the prior written consent of the Surety. Such written consent will not be unreasonably withheld.

9   This Bond shall be governed and construed in accordance with the laws of England and Wales.

_____

IN WITNESS hereof this Bond has been executed as a Deed by the Surety and delivered on the date below:

EXECUTED as a Deed by: _____

for and on behalf of the Surety: _____

EXECUTED as a Deed by: _____

for and on behalf of the Employer: _____

Date: _____

Schedule to Advance Payment Bond
(clause 3(b) of the bond)

Notice of Demand

Date of Notice: _____

Date of Bond: _____

Employer: _____

Surety: _____

The bond has come into effect.

We hereby demand payment of the sum of

£ _____ (amount in words)
which does not exceed the amount of reimbursement for which
the Contractor is in default at the date of this notice.

Address for payment: _____

_____

This Notice is signed by the following persons who are authorised
by the Employer to act for and on his behalf:

Signed by _____

   Name: _____

   Official Position: _____

Signed by _____

   Name: _____

   Official Position: _____

**The above signatures to be authenticated by the Employer's
bankers**

Bond in respect of payment for off-site materials and/or goods

1   THE parties to this Bond are:

(1) _____

whose registered office is at _____

_____

_____ ('the Surety') , and

(2) _____

of _____

_____ ('the Employer').

2   The Employer and _____ ('the Contractor')

have agreed to enter into a building contract for building works

('the Works') at _____ ('the Contract').

3   Subject to the relevant provisions of the Contract as summarised below but with which the Surety shall not at all be concerned:

(a) the Employer has agreed to include in the amount stated as due in Interim Certificates (as defined in the Contract) for payment by the Employer the value of those materials or goods or items pre-fabricated for inclusion in the Works which have been listed by the Employer ('the listed items'), which list has been included as part of the Contract, before their delivery to or adjacent to the Works; and

(b) the Contractor has agreed to insure the listed items against loss or damage for their full value under a policy of insurance protecting the interests of the Employer and the Contractor during the period commencing with the transfer of the property in the items to the Contractor until they are delivered to or adjacent to the Works; and

(c) this Bond shall exclusively relate to the amount paid to the Contractor in respect of the listed items which have not been delivered to or adjacent to the Works.

4   The Employer shall in making any demand provide to the Surety a Notice of Demand in the form of the **Schedule** attached hereto which shall be accepted as conclusive evidence for all purposes under this Bond. The signatures on any such demand must be authenticated by the Employer's bankers.

5   The Surety shall within 5 Business Days after receiving the demand pay to the Employer the sum so demanded. 'Business Day' means the day (other than a Saturday or a Sunday) on which commercial banks are open for business in London.

6   Payments due under this Bond shall be made notwithstanding any dispute between the Employer and the Contractor and whether or not the Employer and the Contractor are or might be under any liability one to the other. Payment by the Surety under this Bond shall be deemed a valid payment for all purposes of this Bond and shall discharge the Surety from liability to the extent of such payment.

7   The Surety consents and agrees that the following actions by the Employer may be made and done without notice to or consent of the Surety and without in any way affecting changing or releasing the Surety from its obligations under this Bond and the liability of the Surety hereunder shall not in any way be affected hereby. The actions are:

(a) waiver by the Employer of any of the terms, provisions, conditions, obligations and agreements of the Contractor or any failure to make demand upon or take action against the Contractor;

(b) any modification or changes to the Contract; and/or

(c) the granting of an extension of time to the Contractor without affecting the terms of clause 9(b) below.

8   The Surety's maximum liability under this Bond shall be *[_____].

9   The obligations of the Surety and under this Bond shall cease upon whichever is the earlier of

(a) the date on which all the listed items have been delivered to or adjacent to the Works as certified in writing to the Surety by the Employer; and

(b) [*longstop date to be given*],

and any claims hereunder must be received by the Surety in writing on or before such earlier date.

10   The Bond is not transferable or assignable without the prior written consent of the Surety. Such written consent will not be unreasonably withheld.

11   This Bond shall be governed and construed in accordance with the laws of England and Wales.

*The value stated in the Contract which the Employer considers will be sufficient to cover him for maximum payments to the Contractor for the listed items that will have been made and not delivered to the site at any one time.

_____

IN WITNESS hereof this Bond has been executed as a Deed by the Surety and delivered on the date below:

EXECUTED as a Deed by: _____

   for and on behalf of the Surety: _____

EXECUTED as a Deed by: _____

   for and on behalf of the Employer: _____

Date: _____

Schedule to Bond
(clause 4 of the bond)

Notice of Demand

Date of Notice: _____

Date of Bond: _____

Employer: _____

Surety: _____

We hereby demand payment of the sum of _____
being the amount stated as due in respect of listed items included
in the amount stated as due in an Interim Certificate(s) for payment
which has been duly made to the Contractor by the Employer but
such listed items have not been delivered to or adjacent to the
Works.

Address for payment: _____

_____

This Notice is signed by the following persons who are authorised
by the Employer to act for and on his behalf:

Signed by _____

Name: _____

Official Position: _____

Signed by _____

Name: _____

Official Position: _____

**The above signatures to be authenticated by the Employer's
bankers**

Annex 2 to the Conditions: Supplemental Provisions for EDI

(clause 1.11)

The following are the Supplemental Provisions for EDI referred to in clause 1.11 of the Conditions.

1   The Parties no later than when there is a binding contract between the Employer and the Contractor shall have entered into the Electronic Data Interchange Agreement identified in the Appendix ('the EDI Agreement'), which shall apply to the exchange of communications under this Contract subject to the following:

    .1   except where expressly provided for in these provisions, nothing contained in the EDI Agreement shall override or modify the application or interpretation of this Contract;

    .2   the types and classes of communication to which the EDI Agreement shall apply ('the Data') and the persons between whom the Data shall be exchanged are as stated in the Contract Documents or as subsequently agreed in writing between the Parties;

    .3   the Adopted Protocol/EDI Message Standards and the User Manual/Technical Annex are as stated in the Contract Documents or as subsequently agreed in writing between the Parties;

        The EDI Association Standard EDI Agreement refers to an Adopted Protocol and User Manual; the European Model EDI Agreement refers to EDI Message Standards and a Technical Annex. Delete whichever is not applicable.

    .4   where the Contract Documents require a type or class of communication to which the EDI Agreement applies to be in writing it shall be validly made if exchanged in accordance with the EDI Agreement except that the following shall not be valid unless in writing in accordance with the relevant provisions of this Contract:

.4 .1 any determination of the employment of the Contractor;

.4 .2 any suspension by the Contractor of the performance of his obligations under this Contract to the Employer;

.4 .3 the Final Certificate;

.4 .4 any invoking by either Party of the procedures applicable under this Contract to the resolution of disputes or differences;

.4 .5 any agreement between the Parties amending the Conditions or these provisions.

2 The procedures applicable under this Contract to the resolution of disputes or differences shall apply to any dispute or difference concerning these provisions or the exchange of any Data under the EDI Agreement and any dispute resolution provisions in the EDI Agreement shall not apply to such disputes or differences.

Annex 3: Bond in lieu of Retention

BOND dated the _____ day of _____ 20 _____

issued by _____

of _____

_____ (hereinafter called 'the Surety')

in favour of _____

of _____

_____ (hereinafter called 'the Employer').

1   By a building contract ('the Contract' between the Employer and

_____

of _____

_____ (hereinafter called the 'Contractor')

the Employer has agreed that he will not exercise his right under the Contract to deduct Retention from amounts included in Interim Certificates provided the Contractor has taken out this Bond in favour of the Employer.

2   The Surety is hereby bound to the Employer in the maximum

aggregate sum of _____ (figures and words)

until the Surety is notified by the Employer in writing of the date of issue of the next Interim Certificate after Practical Completion when the maximum aggregate sum shall be reduced by 50 per cent.

3   The Employer shall, on a demand which complies with the require-
       ments in clause 4, be entitled to receive from the Surety the sum
       therein demanded.

4   Any demand by the Employer under clause 3 shall:

       (i)   be in writing addressed to the Surety at its office at
              . . . . ,
              refer to this Bond, and with the signature(s) therein authen-
              ticated by the Employer's bankers; and

       (ii)   state the amount of the Retention that would have been held
              by the Employer at the date of the demand had Retention
              been deductible; and

       (iii) state the amount demanded, which shall not exceed the
              amount stated pursuant to clause 4(ii), and identify for which
              one or more of the following such amount is demanded:

              (a) the costs actually incurred by the Employer by reason of
                  the failure of the Contractor to comply with the instruc-
                  tions of the Architect under the Contract; and be accom-
                  panied by a statement by the Architect which confirms
                  that this failure by the Contractor has occurred:

              (b) the insurance premiums paid by the Employer pursuant
                  to the Contract because the Contractor has not taken out
                  and/or not maintained any insurance of the building
                  works which he was required under the Contract to take
                  out and/or maintain;

              (c) liquidated and ascertained damages which under the
                  Contract the Contractor is due to pay or allow to the
                  Employer; and be accompanied by a copy of the certifi-
                  cate of the Architect which under the Contract he is
                  required to issue and which certifies that the Contractor
                  has failed to complete the works by the contractual
                  Completion Date;

              (d) any expenses or any direct loss and/or damage caused to
                  the Employer as a result of the determination of the
                  employment of the Contractor by the Employer;

              (e) any costs, other than the amounts referred to in clauses
                  4(iii)(a), (b), (c) and (d), which the Employer has actually

incurred and which, under the Contract, he is entitled to deduct from monies otherwise due or to become due to the Contractor; and identify his entitlement;

and

(iv) incorporate a certification that the Contractor has been given 14 days' written notice of his liability for the amount demanded hereunder by the Employer and that the Contractor has not discharged that liability; and that a copy of this notice has at the same time been sent to the Surety at its office at

_____

Such demand as above shall, for the purposes of this Bond but not further or otherwise, be conclusive evidence (and admissible as such) that the amount demanded is properly due and payable to the Employer by the Contractor.

5   If the Contract is to be assigned or otherwise transferred with the benefit of this Bond, the Employer shall be entitled to assign or transfer this Bond only with the prior written consent of the Surety, such consent not to be unreasonably delayed or withheld.

6   The Surety, in the absence of a prior written demand made, shall be released from its liability under this Bond upon the earliest occurrence of either

(i)   the date of issue under the Contract of the Certificate of Completion of Making Good Defects; or

(ii)   satisfaction of a demand(s) up to the maximum aggregate under the Bond; or

(iii) _____ (insert calendar date).

7   Any demand made hereunder must be received by the Surety accompanied by the documents as required by clause 4 above on or before the earliest occurrence as stated above, when this Bond will terminate and become of no further effect whatsoever.

8   This Bond is subject to English law.

Attestation: to be executed as a deed by the Surety.

Supplemental Provisions (the VAT Agreement)

The Supplemental Provisions are incorporated into the Contract by Clause 15.1.

1   The Employer must pay to the Contractor, in the manner specified, any VAT chargeable by the Commissioners (of Customs and Excise) on the Contractor in respect of the supply of goods and services to the Employer under the Contract.

    (Under any current amendments of Regulation 26 of the Value Added Tax (General) Regulations, 1985.)

    Thus, the Employer must pay the Contractor the appropriate amount in respect of any positively rated items in the Contract, at the stipulated rate(s).

1A   .1   This Clause is an alternative to Clauses 1.1 to 1.2.2 of the VAT Agreement; for Clause 1A to apply, it must be so stated in the Appendix. Clause 1A applies in conjunction with Clauses 1 and 1.3 to 8 of the VAT Agreement.

    If the Contractor fails to give the written notice required under Clause 1A.2 or a notice under Clause 1A.4 is given (by either party), Clause 1A ceases to apply.

    .2   The Contractor must give a written notice to the Employer (copy to Architect):

- of rate of tax chargeable on supply of goods and services (for Interim and Final Certificates),

- prior to 7 days before the date for issue of the first Interim Certificate.

Contractor must inform the Employer by written notice (copy to Architect) of changes in the rate of tax within 7 days of such changes coming into effect.

    .3   VAT, at the appropriate rate(s) according to the Contractor's notice(s), must be shown on each Interim Certificate and, subject to Clause 1.3, on the Final Certificate. Sums of VAT payable must be paid, either way, within the appropriate period of honouring.

.4   The Employer or Contractor may give written notice to the other (copy to Architect) that Clause 1A shall apply no longer – effect from the date of the notice.

1.1   Unless Clause 1A applies, the Contractor must give the Employer a written provisional assessment of the values (less any applicable Retention) of supplies of goods and services included in a payment Certificate which are subject to VAT (i.e. all *except exempt* items).

The timing of such assessment must be:

(a)   not later than the date for the issue of each Interim Certificate

(b)   not later than the date for the issue of the Final Certificate, unless the procedure detailed in Clause 1.3 has been completed.

The assessment must show the supplies:

.1   subject to zero rate of tax (Category (i)), *and*

.2   subject to any rate(s) of tax other than zero (Category (ii)), i.e. positively rated.

In respect of any Category (ii) items, the Contractor must state:

(a) the rate(s) applicable, *and*

(b) the grounds upon which he considers the supplies so chargeable.

.2   .1   On receipt of a written provisional assessment, the Employer must (unless he (may reasonably) objects to the assessment):

(a) calculate the amount of tax due on the basis of the details of the Category (ii) items,

(b) pay that amount to the Contractor within the period for honouring (together with the amount due ordinarily under the Interim Certificate).

.2   If the Employer has reasonable grounds for objection to the provisional assessment, he must so notify the Contractor in writing, specifying the grounds within 3 working days of receipt of the assessment. Within 3 working days of receipt by the Contractor of the Employer's notice, he must reply in writing either:

(a) withdrawing his assessment and thereby releasing the Employer from the obligations to calculate and pay tax in respect of the assessment – Clause 1.2.1, *or*

(b) confirm the assessment. The Contractor then regards any amount received from the Employer in connection with the assessment and associated Certificate as being inclusive of the appropriate VAT (i.e. a gross payment) and issues an authenticated receipt accordingly – Clause 1.4.

.3   .1   If Clause 1A applies, Clause 1.3 applies only if no amount of VAT has been shown on the Final Certificate. As soon as possible after the issue of the Certificate of Completion of Making Good Defects, the Contractor must prepare a final written statement of the respective values of all supplies of goods and services for which certificates have been or will be issued and which are chargeable on the Contractor at

.1   zero rate – Category (i), and

.2   any rate(s) of tax other than zero – Category (ii)

The final statement must be issued to the Employer. As with the interim assessments, in respect of Category (ii) items the Contractor must specify:

(a) the rate(s) applicable, *and*

(b) the grounds upon which he considers the supplies so chargeable, *and,* for the final statement only,

(c) the total amount of tax already received by the Contractor for which receipt(s) have been issued – as Clause 1.4 of these provisions.

.2   The final statement may be issued either before or after the issue of the Final Certificate. Practically, it is probably advantageous to issue the statement after the issue of the Final Certificate.

.3   The Employer must, on receipt of the final statement, calculate the final amount of tax due upon the details of the statement. He must pay any balance of tax (total per statement calculations less tax already paid) to the Contractor within 28 days from receipt of the statement.

.4   If the Employer discovers the amount of tax in accordance with the final statement is less than the amount he has already paid, he must so notify the Contractor. The Contractor then must:

(a) refund the excess to the Employer within 28 days of receipt of the notification, *and*

(b) accompany the refund by a receipt (Clause 1.4 of the provisions) showing the correction of previous receipt(s).

The Contractor must issue receipts upon receiving monies under certificates and the appropriate amount of tax (Clause 1 or 1A).

The receipts must comply with Regulation 12(4) of the VAT (General) Regulations, 1985, including the particulars as required by Regulation 13(1), taking account of any amendments and/or re-enactments.

2.1 The Employer must disregard any set-off in respect of liquidated damages when calculating and paying amounts of VAT due to the Contractor (Clauses 1.2 and 1.3 of these provisions).

.2 The Contractor must likewise ignore contra-charges of liquidated damages by the Employer in the preparation of the final statement.

.3 If Clause 1A applies, the Employer must pay VAT in full, irrespective of the provisions of Clause 24.

3.1 If the Employer disagrees with the Contractor's final statement, he may ask the Contractor to obtain the decision of the Commissioners on the tax properly chargeable. This request must be made before the tax payment (or refund) becomes due.

If the Employer then disagrees with the Commissioners' decision, he must request the Contractor to make such appeals to the Commissioners as he instructs. In such instances the Employer must indemnify the Contractor against all costs and other expenses. (The Contractor also has an option to be secured by the Employer against such costs and expenses.)

The Contractor must account to the Employer for any costs awarded in his favour in any appeals under Clause 3 of the provisions.

.2 If required to do so, before an appeal may proceed, the Employer must pay to the Contractor the full amount of tax alleged to be chargeable.

.3 The balance of tax must be paid by or refunded to the Employer within 28 days of the final adjudication of an appeal or from the date of the Commissioners' decision if no appeal is to be made.

Authenticated receipts must also be issued accordingly under Clause 1.3.4 of these provisions.

4   The Employer is discharged from further liability to pay tax to the Contractor upon settlement, as prescribed, of the amount of the final statement, Commissioners' decision or appeal decision.

The exception is if the Commissioners introduce a correction to the tax charged subsequently. (This is also subject to the prescribed appeal provisions of Clause 3 of these provisions.) Thereupon the Employer must pay the additional amount to the Contractor.

5   Awards of an Arbitrator or Court which vary payments between the parties. Such alterations in payments must also be subject to VAT as applicable.

6   Arbitration is not applicable to VAT assessments by the Commissioners (Clause 3 of these provisions).

7   If the Contractor does not provide a receipt, the Employer is not obliged to make any further payments to the Contractor. This applies only if

.1   The Employer shows that he requires the receipt to validate a claim for credit for tax paid or payable under the Agreement which he is titled to make to the Commissioners, *and*

.2   The Employer has paid tax in accordance with the provisional assessments unless he has sustained a reasonable objection thereto.

8   If the Employer determines the employment of the Contractor (Clause 27.4 of the Conditions), any additional tax which the Employer has had to pay due to the determination may also be set-off by him against any payments to be made to the Contractor (or may be recovered by the Contractor).

Sectional Completion Supplement

The supplement applies to both the Private and Local Authorities Editions, with Quantities.

The Contract Standard Forms do permit partial possession by the Employer – Clause 18. However, that Clause is not intended to be used in such a way that the Contract is, in fact, a phased Contract. In such a case the Sectional Completion Supplement must be used.

The decision is a reflection of the Employer's initial intention – if a phased project, use the Sectional Completion Supplement; if a non-phased project but subsequently the Contractor completes in a 'sectional manner' and the Employer wishes to take possession of the completed section(s), Clause 18 is applicable.

Practice Note 1 is a useful guide to practice and provides a set of amendments to the Contract to facilitate sectional completion.

Contractors must be informed at Tender stage:

(a)   The Works are to be carried out in phased Sections.

(b)   The Employer will take possession of each Section on Partial Completion of the Section.

(c)   The identity of the Sections – by drawings or BQ description.

(d)   The order and phasing of completion.

(e)   Any work common to more than one Section must be a separate Section.

(f)   In respect of each Section: Value, Date for Possession, Date for Completion, Rate of Liquidated Damages for Delay, Defects Liability Period.

The major adaptations of the contract involve the division of the Works into distinct Sections with the values ascribed thereto summing to the Contract Sum. Separate completion provisions apply for each Section but only one Final Certificate is issued at the conclusion of the project.

Clause	Adaptation
1.3	defines Section
2.1	Contractor is obliged 'to carry out and complete the Works by Sections'
18.1.6	Section value is 'the value ascribed to the relevant Section in the Appendix'

214

The Section value must be the total value of that Section obtained from the Contract Bills. The Section values must add up to the Contract Sum and must take into account the apportionment of Preliminaries and similar items priced in the Contract Bills.

Procedure for Sectional Completion:

> Contractor to be given possession of each Section.
>
> Contractor to execute the Sections successively or concurrently in accordance with the Contract.
>
> Liquidated Damages are calculated and paid separately.
>
> A Practical Completion Certificate must be issued for each Section as Practical Completion is achieved.

Practical Completion of a Section:

(a) relieves the Contractor of any duty to insure that Section under Clause 22A

(b) Clause 30.4 operates to realise *approximately* one half of the Retention held in respect of that Section

(c) Clause 17.2 – commences a separate DLP for that Section.

When all Sections are complete, the Architect must so certify – the period for the Final Account and Final Certificate for the whole Works commences at the date of that Certificate (Clause 30.6)

Insurance arrangements under Clause 21.2 – amount of insurance to be maintained in respect of persons and property – must be clarified to determine whether or not any Sections of the Works which have obtained a Certificate of Practical Completion are then covered by such insurance as 'property other than the Works'.

The actual changes in the Contract to effect Sectional Completion involve quite minor changes of clause wordings. The changes are, in reality, quite straight-forward and 'common sense' in nature. The amendments to the Appendix provide the greatest amount of information about the requisite changes and in a concise form.

Contractor's Designed Portion Supplement

See also Practice Note CD/2.

This supplement may be used with both private and Local Authorities Editions with Quantities.

The supplement provides express terms to incorporate design by the Contractor into the contract. Otherwise the Contractor has no express liability for design but is subject to design liability only to inform the Architect of suspected design defects (see *Equitable Debenture Asset Corporation Ltd* v. *William Moss & Others* (1984)).

Amendments to be made to the Standard Form are:

(a)   replacement of the Recitals and Article 1 by amended versions

(b)   modifications to the Conditions

(c)   additional of a Supplementary Appendix

(d)   addition of the words 'With Contractor's Designed Portion Supplement' at the top of the endorsement of the outside back cover of the Contract Form.

There are also three additional Contract Documents which must be signed by or on behalf of the parties, they are:

(a)   the Employer's Requirements for the Contractor's Designed Portion (CDP)

(b)   the Contractor's Proposals for the CDP, and

(c)   the Analysis of the portion of the Contract Sum to which the CDP relates.

The three additional Contract Documents must be produced as part of the tendering processes – the Employer's Requirements, by or on behalf of the Employer; the Contractor's Proposals and the Analysis, by the Contractor.

The Architect retains overall responsibility for design and is responsible for the integration of the CDP into the Works and to issue directions regarding such integration; the Architect's powers in this regard are contained in Article 1 and in Clause 2.1.3.

The Second Recital identifies the CDP as part of the Works (the latter being denoted in the First Recital).

The Third Recital identifies who has prepared the drawings etc., including the Employer's Requirements.

The Fourth Recital requires that the Contractor has submitted the Contractor's Proposals' for the design and construction of the CDP and 'the Analysis' of the part of the Contract Sum relating to the CDP.

Note: Contents and format of the Analysis are at the discretion of the Contractor unless requirements therefore, such as those outlined in Practice Note CD/2, are provided in the tender documents – particularly the Bills of Quantities.

The Seventh Recital stipulates that the Employer has examined the Contractor's Proposals in the context of the Conditions and is satisfied that the proposals appear to meet the Employer's Requirements.

Note: The Recital indicates acceptance in principle of the proposals but seeks to ensure that the Contractor retains responsibility for their actually meeting the Employer's Requirements.

Other Recitals are almost identical in content to those of JCT 80 but incorporate appropriate amendments to allow for the CDP. There are eight Recitals in the CDP Supplement.

Article 1, Contractor's Obligations incorporate provision for the Contractor to complete the design of the CDP and to do so as the Architect may direct in order that the design of the CDP may be integrated with the design for the Works. The design obligation is in addition to the obligation for the Contractor to carry out and to complete the Works as per the Contract Documents, including the modifications thereto specified within the CDP Supplement.

The remainder of the CDP Supplement details modifications to be made to the Conditions. Many of the modifications are straight-forward and are necessary to incorporate the additional documents, definitions, etc.

Some of the more important amendments are noted below:

Clause	Adaptation
2.1.2	Extends the scope of Contract Documents.
2.1.3	Contractor to design the CDP and to incorporate all specifications etc. not noted in the Employer's Requirements but which are necessary to achieve completion of the Works. Architect may give directions re. design of CDP for integration with the total design for the Works.
2.1.4	Where the Architect is required to approve standards of materials or work, the standards are to be to the Architect's reasonable satisfaction.

2.2.2.3   Errors of description, quantity or omission from the Contractor's proposals and the Analysis must be corrected with no change to the Contract Sum; this includes errors, etc. which necessitate an AI constituting a Variation for their correction.

2.4.1   Discrepancies, divergencies between Contractor's Proposals, Analysis and other design documents prepared by the Contractor – Contractor to notify the Architect of the discrepancy/divergence and, as soon as possible, to provide a statement of proposed amendments to overcome the problem. Once the Architect has received the statement, the obligation for the Architect to issue instructions under Clause 2.3 comes into effect.

2.4.2   AIs given following Clause 2.4.1 shall not occasion any addition to the Contract Sum.

2.5   Provision of supplementary documents, levels, etc. to amplify the Contractor's proposals by the Contractor to the Architect without change.

2.6.1   Contractor assumes design liability for the CDP to the same level as an Architect or appropriate professional designer holding themselves out to be competent to execute the design.

           Note: The standard of care which the Contractor must assume is that of an expert. *See Bolam* v. *Friern Barnet Hospital* (1957) and *Greaves* v. *Baynham Miekle* (1975).

2.6.2   Passes on to the Contractor responsibility under Clause 2.6.1 for compliance with Section 2(i) of the Defective Premises Act, 1972 where the CDP concerns work in connection with a dwelling or dwellings and the Employer's Requirements refer to such liability.

2.6.3   Where the Contract does not involve the Contractor working in dwellings under the Defective Premises Act, 1972, the Contractor's liability for other parties' losses arising out of the Contractor's failure under Clause 2.6.1 is limited to the amount if any as stated in the Supplementary Appendix. This Clause has no bearing upon the damages provisions under Clause 24.1

2.6.4   Contractor's design includes designs prepared by others on behalf of the Contractor.

2.7   If the Contractor believes that compliance with an AI or any direction given under Clause 2.1.3 injuriously affects the efficacy of the design of the CDP, the Contractor has 7 days from receipt of the AI or direction to so notify the Architect in writing. The notice must specify the injurious effect(s). The AI or direction will not take effect unless confirmed by the Architect.

2.8   The Architect must notify to the Contractor any items which appear to the Architect to be defects in design under Clause 2.6.1 (i.e. design defects in the Contractor's Design). Despite any such notices, the Contractor retains full responsibility for the Contractor's Design.

2.9   No extension of time may be awarded under Clause 25.3 nor any loss and/or expenses paid under Clause 26.1 and 28.1.3 in respect of:

     .1   errors, divergencies etc. in Contractor's Proposals or supplementary documents provided under Clause 2.5

     .2   late provision by the Contractor of drawings etc. concerning the CDP as per Clause 2.5.2

     .3   non receipt in due time by the Architect of drawings, etc. concerning the CDP from the Contractor for which the Architect, specifically applied in writing on a date which having regard to the Completion Date was neither unreasonably distant from, nor unreasonably close to, the date on which it is necessary for him to receive the same.

5.9   Prior to the commencement of the DLP, the Contractor shall provide free to the Employer drawings, etc., concerning the CDP re maintenance and operation of that Portion as specified in the Contract Documents or as the Employer reasonably may require.

     Note: It would be advisable for the Contractor to provide for giving the Employer a full set of as-built drawings, etc. and any available maintenance documents (e.g. maintenance manuals of plant).

6.1.6   If either the Contractor or the Architect finds any divergence between the Contractor's Proposals, etc. (Clauses 2.3.5, to 2.3.8) and the Statutory Requirements, he must give the other written notice specifying the divergence. The Contractor must give the Architect written notice of his proposed amendment(s) to remove the divergence; the Architect shall issue instructions to remove the divergence and the Contractor must comply with such AIs free of charge. However, if the cause of the divergence is changes in the Statutory Requirements which occurred after the Date of Tender (Clause 6.11.8) and necessitate a change to the CDP, such change shall be treated as a Variation under Clause 13.2.

6.1.7   Emergency work to comply with Statutory Requirements – Contractor to execute only what is reasonably necessary to secure compliance and to inform the Architect of what is being done.

8.1   As far as possible, materials goods and work in the CDP to comply with Employer's Requirements. Substitutions require the Architect's written consent. This is in addition to the materials etc. so far as procurable complying with the Contract Bills where not covered by the CDP.

13.2   Architect may issue an instruction requiring a Variation. No Variation . . . shall vitiate this Contract (See Clause 13 of main text). Architect may sanction in writing any Variation made by the Contractor which is not covered by an AI. Any AI regarding the content of the CDP acts as a statement of amendment to the Employer's Requirements.

13.4.1   Valuation of work done under AIs re expenditure of provisional sums:

           by QS as Clause 13.5, but
           by QS as Clause 13.8 in respect of CDP Works.

13.8   Valuation rules for Variations to CDP works. Similar rules to valuation of Variations to the main works but:

           (a) based on the Analysis instead of the Contract Bills, and

           (b) to incorporate adjustment for the addition/omission of design work.

19.2.2   Written consent of the Employer must be obtained for the Contractor to sub-let any design for the CDP. Consent may not be unreasonably withheld. Contractor retains full responsibility towards the Employer under Clause 2.6.

30.10   Usually no certificate of the Architect will be conclusive evidence that:

      .1   any works, materials or goods to which it relates; or

      .2   any design to be prepared and completed by the Contractor for the Contractor's Designed Portion, are in accordance with this Contract.

The Supplementary Appendix identifies:

The Employer's Requirements
The Contractor's Proposals
The Analysis

Clause 2.6.2: whether the scheme is approved under S.2(1) of the Defective Premises Act, 1972.

Clause 2.6: Limit of Contractor's Liability for loss of use, etc.:

Clause 2.6.3 does not apply or

Clause 2.6.3 applies, with a limit of £___ (to be inserted, if applicable)

The bulk of the remaining amendments to incorporate the Contractor's Designed Portion Supplement concern modifications to Clause 40 to facilitate operation of the NEDO price adjustment formulae.

Approximate Quantities Form

See also Practice Note 7.

The Edition is based upon and closely follows the provisions of the With Quantities Form. The main differences are noted below. As the Form uses Approximate Quantities, complete remeasurement of the Works in their 'as executed' form is required.

'Bills of Approximate Quantities' are used throughout, where appropriate, in place of 'Bills of Quantities'.

Recitals: The Bills of Approximate Quantities provide a reasonably accurate forecast of the quantities of work to be executed, the total of the prices in which form the Tender Price.

Article 2:	The total sum to be paid to the Contractor is the Ascertained Final Sum.
Clause 2.3:	Quantity divergencies, etc. are dealt with by remeasurement under Clause 14.
Clause 13:	Corresponds to Clause 14 of the With Quantities Form.
Clause 14:	Corresponds to Clause 13 of the With Quantities Form. Variations of quantity of work do not apply except where the quantity was not accurately forecast in the Bills – it is submitted that the concept of reasonableness will be applicable to determine what constitutes an accurate forecast.
Clause 25.4.13:	An extra Relevant Event, where the quantity of work was not forecast accurately.
Clause 30.1.2:	Valuations are required for each Interim Payment.
Clause 30.6:	Notably different from the With Quantities Form. The approach is to build up the final account from 'nothing' by measurement and valuation of the work done and claims applicable under the terms of the Contract.
Clause 38:	Not available for use under the Approximate Quantities Form.
Clause 40:	Use of Work Groups or a Single Index is inappropriate.

Without Quantities Forms

The Without Quantities Forms are extremely similar to their With Quantities counterparts.

Most of the changes are obviously brought about by the absence of a Bill of Quantities and the presence, in its place, of a Specification and a Schedule of Rates.

Serveral Clauses are amended solely by the substitution of reference to the 'Specification and/or Schedule of Rates' for those to the 'Contract Bills'.

The more significant alterations are considered below.

Article 4:   This is now split into two alternatives:

Article 4A: a reproduction of Article 4, to be used where a Quantity Surveyor will be employed.

Article 4B: a slight change of working from Article 4, to be used where the functions of the Quantity Surveyor are to be carried out by someone who is not a QS.

Clause 1.3:   The definition of Contract Bills is omitted. Definitions are included for

    (a)   Schedule of Rates – Clause 5.3.1.3

    (b)   Specification – First Recital

Clause 2.2.2: This Clause replaces that in the With Quantities Form which is subdivided into two sections.

The Clause states that if any errors or omissions exist in the Contract Drawings and/or Specification they will not vitiate the Contract (but see qualifying notes under the With Quantities form) but must be corrected under the rules of Clause 13.2, as if they were Variations due to Architect's Instructions.

Clause 13.5:   This Clause is largely similar to its With Quantities counterpart, except for the following:

Clause 13.5.3 is omitted and the substitution is made:

Clause 13.5.3 now requires that in the valuation of any Variation (Clause 13.5.1 and 13.5.2) appropriate allowances must be included for any consequent changes in Preliminaries (as defined by SMM7).

Student's Revision Guide

This section provides a list of the contents of the Contract. The information is tabulated in four columns:

Column 1 – Article or Clause number

Column 2 – Article or Clause title – indicating the provisions

Column 3 – Note of the persons primarily concerned with the operation of the provisions

Column 4 – Brief summary of the main provisions.

It should be noted that everyone connected with the operation of the Contract should be aware of all the provisions. Particularly the Employer and Contractor will be affected by every provision. The purpose of column 3, therefore, is to denote those persons most usually and directly affected by the operation of the part of the Contract.

The following abbreviations have been used:

Arch.	Architect
BI	Building Inspector
Contr	Contractor
CoW	Clerk of Works
DS/C	Domestic Sub-Contractor
Empl.	Employer
Govt	Government
LA	Local Authority
NS/C	Nominated Sub-Contractor
NSup	Nominated Supplier
QS	Quantity Surveyor
RIBA	Royal Institute of British Architects
SU	Statutory Undertaker.

Ref. No	Subject	Persons primarily concerned	Major contents/comments
ARTICLES OF AGREEMENT			
1	Contractor's obligations	Contractor	'Blanket' provision – execute prescribed work
2	Contract Sum		Subject only to express modifications – fluctuations
3	Architect	Architect	Employer's agent for express purposes – condition
4	Quantity Surveyor	QS	Must execute variations. Must execute interim valuations if Clause 40 applies.
5	Dispute or difference	Contr. Empl. Arch. QS CoW RIBA Adjudicator	Adjudication in compliance with HGCR
6	CDM Regs	Planning Supervisor Principal Contractor	Identifies Planning Supervisor / Identifies Principal Contractor
7A	Arbitration	Arbitrator	Primarily after Practical Completion – Arbitration
7B	Legal proceedings		Primarily after Practical Completion – Litigation
Conditions: Part 1 – General			
1	Interpretation, definitions etc.	All	
1.1	Method of reference to clauses	All	
1.2	Articles, etc. to be read as a whole	All	As with any legal document
1.3	Definitions		
1.4	Contractor's responsibility	Empl. Arch. CoW Contr.	Contractor responsible for executing work
2	Contractor's obligations		
2.1	Contract documents	All	Drawings, Bills, Articles, Conditions, Appendix
2.2.1	Contract Bills – relation to Articles, Conditions and Appendix	QS	Bills do not override
2.2.2	Preparation of Contract Bills – errors in preparation etc.	QS	Corrected as a Variation
2.3 and 2.4	Discrepancies in or divergences between documents	Arch. QS Contr.	AI required – Contractor's request

Ref. No	Subject	Persons primarily concerned	Major contents/comments
3	Contract Sum – additions or deductions – adjustments – Interim Certificates	Contr. Arch. QS	Adjusted a.s.a.p.
4	Architect's instructions		
4.1	Compliance with Architect's instructions	Contr. Arch.	Immediate unless objection; express authority for AI.
4.2	Provisions empowering instructions	Arch.	Architect may specify provision allowing AI.
4.3.1	Instructions to be in writing	Arch.	
4.3.2	Procedure if instructions given otherwise than in writing	Arch.	Confirmation by Contractor or Architect
5	Contract documents – other documents – issue of certificates	Contr.	
5.1	Custody of Contract Bills and Contract Drawings	Arch. QS Contr.	Architect or QS, Contractor may inspect
5.2	Copies of documents	Contr. Arch.	Contractor's entitlement
5.3	Descriptive schedules, etc. master programme of Contractor	Contr. Arch.	Bills should specify any required master programme
5.4	Drawing or details	Arch. Contr.	2 free copies for Contractor
5.5	Availability of certain documents	Contr.	
5.6	Return of drawings etc.	Arch. Contr.	Property of Architect
5.7	Limits to use of documents	All	For the project only
5.8	Issue of Architect's certificates	Contr. (QS)	
5.9	Supply of as-built drawings, etc. Performance Specified Work	Contr. Empl.	Contractor to supply (free) as-built drawings etc of Performance Specified Work.
6	Statutory obligations, notices fees and charges	BI, etc.	
6.1.1 to .5	Statutory Requirements	Contr. Arch.	Contractor must comply; discrepancy AI? Emergencies
.6	Divergence – Statutory Requirements and the Contractor's Statement	Contr. Arch.	Written notice; Contractor proposes amendment, AI needed

Ref. No	Subject	Persons primarily concerned	Major contents/comments
.7	Change in Statutory Requirements after Base Date	Contr. Arch.	Treated as AI of Variation
6.2	Fees or charges	Contr. QS Empl.	Usually Contractor's responsibility
6.3	Exclusion of provisions on Domestic and Nominated Sub-Contractors	LA SU	Work by LA/Statutory Undertaker, i.e. statutory obligations
7	Levels and setting out of the Works	Arch. Contr.	Architect to provide information, Contractor to execute
8	Work, materials and goods		
8.1.1 to .3	Kinds and standards, etc.	QS Contr. Arch.	As in BQ
.4	Substitution of materials or goods – Performance Specified Work	Contr. Arch.	May substitute with Architect's written consent.
8.2.1	Vouchers – materials and goods	Contr. Arch.	Of compliance – Contractor to provide if Architect requests.
.2	Executed work	Arch.	Architect to express any dissatisfaction within reasonable time.
8.3	Inspection – tests	Arch. Contr. (QS)	AI – cost to Employer if OK; Contractor if defective
8.4	Powers of Architect – work not in accordance with the Contract	All	AI for removal; accept and deduction from Contract Sum; Variations; extent of defects?
8.5	Powers of Architect – non-compliance with Clause 8.1.3	All	AIs of Variation, no adjustment of Contract Sum, no extension of time, etc.
8.6	Exclusion from the Works of persons employed thereon	Arch. Contr.	AI transfer, not sacking
9	Royalties and Patent Rights		
9.1	Treatment of royalties etc. – indemnity to Employer	Contr. Empl. (QS)	By Contractor
9.2	Architect's instructions – treatment of royalties, etc.	Arch. Contr. Empl. QS	Add to contract sum
10	Person-in-charge	Contr.	Always on site – Contractor's Agent

Ref. No	Subject	Persons primarily concerned	Major contents/comments
11	Access for Architect to the Works	Contr.	Including QS, etc. – working hours
12	Clerk of Works	Empl. Arch. CoW	Appointed by Employer, briefed by Arch., issues written directions
13	Variations and provisional sums		
13.1	Definition of Variation	All	'All embracing' – limited by size
13.2	Instructions requiring a Variation	Arch. Contr. (QS)	AI or confirmation
13.3	Instructions on provisional sums	Arch. Contr. QS	AI required
13.4	Valuation of Variations and provisional sum work	QS	By QS
13.4 Alt A	Contractor's Price Statement	Contr.	Contractor submits Price Statement
13.4 Alt B	QS Valuation	QS	QS values
13.5	Valuation rules	QS Contr. Arch.	As BQ, pro-rata, fair, daywork, claim?
13.6	Contractor's right to be present at measurement	QS Contr.	
13.7	Valuations – addition to or deduction from Contract Sum		
13A	Contractor's Quotation Contract Sum	Contr.	Price in advance for AI
14	Contract Sum	All	BQ
14.1	Quality and Quantity of work included in Contract Sum		
14.2	Contract Sum – only adjusted under the Conditions – errors in computation	Empl. Contr. QS	Arithmetic errors – deemed accepted by parties
15	Value added tax – supplemental provisions	Customs & Excise	
15.1	Definitions – VAT Agreement	Finance Act 1972	
15.2	Contract Sum – exclusive of VAT	All	

Ref. No	Subject	Persons primarily concerned	Major contents/comments
15.3	Possible exemption from VAT		Input tax adjustment
16	Materials and goods unfixed or off-site		Transfer of title provisions
16.1	Unfixed materials and goods – on site	All	Remove only for incorporation or AI required
16.2	Unfixed materials and goods – off site	All	If included in a Certificate – Contractor's responsibilities pass on to supplier etc.
17	Practical completion and Defects liability		Complete for practical purposes, NOT 'substantial completion'
17.1	Certificate of Practical Completion	Arch. Empl. Contr. QS	Date of Practical Completion as Certificate
17.2	Defects, shrinkages or other faults	Arch. Contr.	Schedule – make good by Contractor at his cost.
17.3	Defects etc. – Architect's instructions	Arch. Contr.	AI to make good prior to schedule
17.4	Cert. of Completion of Making Good Defects	Arch. Contr. (QS Empl.)	Architect's opinion
17.5	Damage by frost	Contr.	Contractor's responsibility – frost prior to Practical Completion
18	Partial possession by Employer		Architect to certify parts and date of possession by Employer
18.1	Employer's wish – Contractor's consent	Empl. Arch. Contr.	
18.1.1	Practical Completion – relevant part	All	As total contract provisions but to the part
18.1.2	Defects etc. – relevant part	Arch. Contr.	As total contract provisions but to the part
18.1.3	Insurance – relevant part	Empl. Contr.	Employer's responsibility
18.1.4	Liquidated damages – relevant part	Empl. Contr.	Pro-rata adjustment
19	Assignment and Sub-Contracts		
19.1	Assignment	Empl. Contr.	Not without consent of other party (written)

Ref. No	Subject	Persons primarily concerned	Major contents/comments
19.2	Sub-letting – Domestic Sub-Contractors – Architect's consent	Arch. Contr.	Min. 3 persons
19.3	Sub-letting – list in Contract Bills	Arch. Contr. QS	
19.4	Sub-letting – determination of employment of Domestic Sub-Contractors	Contr. DS/C	Immediately main Contractor's employment determined Property in materials on site, placed thereon by by Domestic Sub-Contractors
19.5	Nominated Sub-Contractors	Contr. NS/C Arch. QS	Part 2, nomination must occur, Contractor not obliged to execute the work, etc.
20	Injury to persons and property and Employer's indemnity		
20.1	Liability to Contractor – personal injury or death – indemnity to Employer	Contr. All	Appendix-amount, unless Employer's (etc.) negligence, etc.
20.2	Liability of Contractor – injury or damage to property – indemnity to Employer	Contr. All	Appendix-amount due to Contractor's (etc.) negligence etc.
20.3	Injury or damage to property – exclusion of the Works and Site Materials	Contr. All	
21	Insurance against injury to persons or property	All	
21.1	Insurance – personal injury or death – injury or damage to property	Contr. Arch.	Minimum cover as Appendix. Evidence of insurance – inspection by Employer via Architect
21.2	Insurance liability etc. of Employer	Empl. Contr. Arch.	Appendix – whether cover required – Arch. to instruct Joint Names Policy – Indemnity amount as Appendix note on exceptions
21.3	Excepted risks	Empl.	Assumed by Employer
22	Insurance of the Works	All	

Ref. No	Subject	Persons primarily concerned	Major contents/comments
22.1	Insurance of the Works – alternative clauses	Empl. Contr. Arch. QS	22A or 22B or 22C Joint Names Policy for All Risks Insurance 22A, 22B Joint Names Policy re: Specified Perils – 22C
22.2	Definitions	All	All Risks Insurance, Site Materials Special Provisions for projects in Northern Ireland
22.3	Nominated and Domestic Sub-Contractors – benefit of Joint Names Policies – Specified Perils	All	NS/Cs & DS/Cs insured under Joint Names Policy – no subrogation
22A	Erection of new buildings – All Risks Insurance of the works by the Contractor		
22A.1	New buildings – Contractor to take out and maintain a Joint Names Policy for All Risks Insurance	Empl. Contr. Arch. QS	Note end date for cover – full reinstatement value Adequacy of cover – full reinstatement value Not for Contr.'s plant, huts, etc.
22A.3	Use of annual policy maintained by Contractor – alternative to use of Clause 22A.2	Contr.	'All risks' policy; check scope of cover – Clause 22.2
22A.4	Loss or damage to Works – insurance claims – contractor's obligations – use of insurance monies	All	Contr. – give written notice of loss/damage Contr. to repair, etc. Insurance monies paid to Empl.; only these available to pay Contr. for repairs etc.
22B	Erection of new buildings All Risks Insurance of the Works by the Employer	All	
22B.1	New buildings – Employer to take out and maintain a Joint Names Policy for All Risks Insurance	Empl. Contr. Arch.	New building

Ref. No	Subject	Persons primarily concerned	Major contents/comments
22B.2	Failure of Employer to insure – rights of Contractor	Empl. Contr. Arch. QS	Employer to show evidence of insurance Contractor may insure and charge if Employer defaults.
22B.3	Loss or damage to Works – insurance claims – Contractor's obligations – payment by Employer	Empl. Contr. Arch. QS	As Clause 22A.3 but repairs etc. paid for as a Variation under Clause 13.2
22C	Insurance of existing structures Insurance of Works in or extensions to existing structures	All	
22C.1	Existing structures and contents – Specified Perils – Employer to take out and maintain Joint Names Policy	Empl. Contr. Arch. QS	Insure for full cost of reinstatement etc. All monies paid to Employer
22C.2	Works in or extensions to existing structures – All Risks Insurance – Employer to take out and maintain Joint Names Policy	Empl. Contr. Arch.	
22C.3	Failure of Employer to insure – rights of Contractor	Empl. Contr. Arch. QS	Employer to produce evidence of insurance Contractor may insure and charge if Employer defaults
22C.4	Loss or damage to Works – Insurance claims – Contractor's obligations – payment by Employer	All	Contractor – written notice of loss/damage to Employer – Insurance monies paid to Employer Contractor to repair etc.? If so paid as Variation.
22D	Insurance for Employer's loss of liquidated damages – Clause 25.4.3	Empl. Contr. Arch.	Optional – see Appendix. If required Architect to inform Contractor to obtain quotation. AI re Employer's acceptance. Period as Appendix or extension of time.

Ref. No	Subject	Persons primarily concerned	Major contents/comments
22FC	Joint Fire Code	Empl. Contr. Arch.	Application of Joint Fire Code
23	Date of Possession, completion and Postponement	All	
23.1	Date of Possession – progress to Completion Date	Contr. Arch. Empl.	Appendix. Completion by Completion Date
23.2	Architect's instructions – postponement	Arch. Contr.	Of any work
23.3	Possession by Contractor – use of occupation by Employer	All	Contractor's consent required. Insurance cover to be unaffected – consult insurers
24	Damages for non-completion	Arch. Contr. QS	Appendix – genuine pre-estimate of Employer's loss
24.1	Certificate of Architect	Arch. QS Contr.	Required as pre-requisite to payment
24.2	Payment or allowance of liquidated damages	Empl.	Extended completion date? Any partial possession?
25	Extension of Time	All	
25.1	Interpretation of delay		
25.2	Notice by Contractor of delay to progress	Contr. Arch.	Pre-requisite for each delay – Contractor to write to Architect? period
25.3	Fixing Completion Date	Arch. (Contr. QS)	Architect – never earlier than original date in Contract
25.4	Relevant Events	Contr. Arch.	List, Contractor to specify, Architect to evaluate – 12 weeks
26	Loss and expense caused by matters materially affecting regular progress of the Works		Architect (or QS) to evaluate
26.1	Matters materially affecting regular progress of the Works – direct loss and/or expense	Contr. Arch. QS	Contractor's written application – as soon as apparent
26.2	List of matters	Contr. (Arch. QS Empl.)	Less than Clause 1.25; Employer's responsibility and control

Ref. No	Subject	Persons primarily concerned	Major contents/comments
26.3	Relevance of certain extensions of Completion Date	Arch.	Also entitled to claim
26.4	Nominated Sub-Contractors – matters materially affecting regular progress of the Sub-Contract Works – direct loss and/or expense	NS/C, Contr. Arch. QS	
26.5	Amounts ascertained – added to Contract Sum	All	
26.6	Reservation of rights and remedies of Contractor	Contr.	'Without prejudice'
27	Determination by Employer		
27.1	Notices under clause 27	Contr. Arch.	Of Contractor's employment In writing; actual delivery, registered post or recorded delivery.
27.2	Default by Contractor	Contr. Empl. Arch.	Before Practical Completion.
27.3	Insolvency of Contractor	Contr. Empl. Arch.	Unless for reconstruction or amalgamation
27.4	Corruption	All	
27.5	Insolvency of Contractor – option to Employer	All	Agree for completion with 'Contractor'?
27.6	Consequence of determination under clauses 27.2 to 27.4	All	Others to complete and set-off?
27.7	Employer decides not to complete the Works	All	Notice to Contractor within 6 months.
27.8	Other rights and remedies	All	Without prejudice.
28.1	Notices under Clause 28	Contr. Arch. Empl.	In writing; actual delivery, registered post or recorded delivery
28.2	Default by Employer – suspension of uncompleted Works	All	Non-payment (incl. VAT), interferes with certificate, suspends execution of Work – Appendix
28.3	Insolvency of Employer	All	Employer to notify Contractor

Ref. No	Subject	Persons primarily concerned	Major contents/comments
28.4	Consequences of determination under clause 28.2 or 28.3	All	Settlement of account
28.5	Other rights and remedies	All	Without prejudice
28A	Determination by Employer or Contractor		
28A.1	Grounds for determination of the employment of the Contractor	Empl. Contr.	Before Practical Completion. Force majeure, etc.
28A.2 to 28A.6	Consequences of determination under clause 28A.1.1 – Clauses 28A.3 to 28A.6	All	Remove huts etc. Provide documents and settle accounts.
28A.7	Amounts attributable to any Nominated Sub-Contractors	Empl. Contr. NS/C	Employer to give written notices.
29	Works by Employer or persons employed or engaged by Employer		'Artists and Tradesmen'
29.1	Information in Contract Bills	QS Contr. Empl.	Not part of Contract
29.2	Information not in Contract Bills	Contr. Empl.	
30	Certificates and payments		
30.1	Interim Certificates and valuations	Arch. QS Contr.	Monthly (usually). 14 days to honour. By Architect.
30.2	Ascertainment of amounts due in Interim Certificates	Arch. QS	Gross
30.3	Off-site materials or goods	Arch. Contr. QS	Include if listed items
30.4	Retention – rules for ascertainment	Arch. QS Contr.	5% or 3% usually
30.5	Rules on treatment of Retention	Arch. QS Contr.	
30.6.1	Final adjustment of Contract Sum documents from Contractor – final Valuation under Clause 13	Arch. QS Contr.	

Ref. No	Subject	Persons primarily concerned	Major contents/comments
30.6.2	Items included in adjustment of Contract Sum	Arch. QS Contr.	
30.7	Interim Certificate – final adjustment or ascertainment of nominated Sub-Contract sums	NS/C, Arch. QS	
30.8	Issue of Final Certificate	Arch. QS Contr.	3 months from end DLP, Completion of m.g., defects etc.
30.9	Effect of Final Certificate	All	Evidence of compliance
30.10	Effect of certificates other than Final Certificate	All	
31	Statutory tax deduction scheme	QS	
31.1	Definitions		
31.2	Whether Employer a 'contractor'	Empl. Contr.	
31.3	Provision of evidence – tax certificate	Contr.	Type of certificate – inspection etc.
31.4.1	Uncertified Contractor obtains tax certificate	Contr. Empl.	Contractor to inform Employer
31.4.2	Expiry of tax certificate	Contr. Empl.	Contractor to inform Employer
31.4.3	Expiry of tax certificate	Contr. Empl.	Contractor to inform Employer
31.5	Vouchers	Contr.	Essential – payments liability etc.
31.6	Statutory deduction – direct cost of materials	Contr. Empl.	Applies to labour content only
31.7	Correction of errors	Empl. Contr.	
31.8	Relation to other clauses		
31.9	Disputes or differences	Empl. Arch. Contr. Arbitrator	
32	(Number not used)		
33	(Number not used)		
34	Antiquities	Arch. CoW	

Ref. No	Subject	Persons primarily concerned	Major contents/comments
34.1	Effect of find of antiquities	Contr.	Stop adjacent work. Contractor to inform Architect or CoW
34.2	Architect's instructions on antiquities found	Arch.	
34.3	Direct loss and/or expense	Arch. QS Contr.	Architect (or QS) to ascertain
CONDITIONS: PART 2: NOMINATED SUB-CONTRACTORS AND NOMINATED SUPPLIERS			
35	Nominated Sub-Contractors General	NS/C	
35.1	Definition of a Nominated Sub-Contractor	QS Contr.	PC Sum in BQ or by AI re provisional sum
35.2	Contractor's tender for works otherwise reserved for a Nominated Sub-Contractor	Contr.	Contractor may be a NS/C
35.3	Procedure for nomination of a Sub-Contractor		
35.4	Documents relating to Nominated Sub-Contractors Procedure for nomination of a Sub-Contractor	All	NSC/T, NSC/A, NSC/C, NSC/W, NSC/N.
35.5	Contractor's right of reasonable objection	Contr.	a.s.a.p. – within 7 days of AI
35.6	Architect's instruction on Nomination NSC/N – documents accompanying the instruction – copy of instruction to sub-contractor	Arch. Contr. NS/C	Nomination AI
35.7	Contractor's obligations on receipt of Architect's instruction	Contr. NS/C. Arch.	Agree NSC/T Part 3.

Ref. No	Subject	Persons primarily concerned	Major contents/comments
35.8	Non-compliance with clause 35.7 – Contractor's obligation to notify Architect	Contr. Arch.	Non Agreement of NSC/T Part 3 – inform Architect
35.9	Architect's duty on receipt of any notice under clause 35.8	Arch. Contr.	New date for agreement; affirm/revise nomination
35.10	(Number not used)		
35.11	(Number not used)		
35.12	(Number not used)		
35.13.1 and .2	Architect – direction as to interim payment for Nominated Sub-Contractor	Arch. Contr. NSC/C	Directions re payments
.3 to .5	Direct payment of Nominated Sub-Contractor	Arch. Contr. Arch. Contr.	Proof of discharge of payments
.6	Agreement NSC/W – pre-nomination payments to Nominated Sub-Contractor by Employer	Empl. NS/C	Direct payments to NS/C
35.14	Extension of period or periods for completion of nominated sub-contract works	NS/C. Contr. Arch.	Contractor grants with Architect's prior written approval.
35.15	Failure to complete nominated sub-contract works	NS/C. Contr. Arch.	Architect may certify non-completion.
35.16	Practical completion of nominated sub-contract works	Arch. NS/C Contr.	Architect to certify Practical Completion.
35.17 to.19	Early final payment of Nominated Sub-Contractors	NS/C. Empl. Arch. Contr.	
35.18	Defects in nominated sub-contract works after final payment of Nominated Sub-Contractor – before issue of Final Certificate	Contr. NS/C Arch. Empl.	

Ref. No	Subject	Persons primarily concerned	Major contents/comments
35.19	Final payment – saving provisions	Empl. NS/C	Contractor's responsibility
35.20	Position of Employer in relation to Nominated Sub-Contractor	Empl.	See NSC/W
35.21	Clause 2.1 of Agreement NSC/W – position of Contractor	Contr.	Limits Contractor's responsibility to Employer
35.22	Restrictions in contracts of sale etc. – limitation of liability of Nominated Sub-Contractors	NS/C. Contr. Empl.	Limits of liability pass to Employer via Contractor
35.23	(Number not used)		
35.24	Circumstances where re-nomination necessary	Contr. Arch. NS/C	Reasonable time
35.25 and 35.26	Determination or determination of employment of Nominated Sub-Contractor – Architect's instructions	Contr. NS/C Arch.	Contractor determines with Architect's prior, written agreement.
NOMINATED SUPPLIERS		N Sup.	
36.1	Definition of Nominated Supplier	Arch. QS	PC sum or single source
36.2	Architect's instructions	Arch. Contr. QS	To nominate
36.3	Ascertainment of costs to be set against prime cost sum	Arch. QS Contr. N Sup.	Architect's opinion
36.4	Sale contract provisions – Architect's right to nominate supplier	Arch. Contr.	
36.5	Contract of sale – limitation or exclusion of liability	Arch. Contr. N. Sup. Empl.	Architect's written approval – liability of Contractor to Employer so limited.
CONDITIONS: PART 3: FLUCTUATIONS			
37	Choice of fluctuation provisions – entry in Appendix	Empl. QS Arch.	Clause 38 unless another identified

Note: Clauses 38, 39 and 40 are published separately

38 CONTRIBUTION, LEVY AND TAX FLUCTUATIONS

Ref. No	Subject	Persons primarily concerned	Major contents/comments
38.1.1	Deemed calculation of Contract Sum types and rates of contribution, etc.		
38.1.2	Increases or decreases in rates of contribution, etc. – payment or allowance		
38.1.3 & .4	Persons employed on site other than 'workpeople'		Craft operative rate fluctuations applicable
38.1.5 to .7	Refunds and premiums		Deemed not contracted out
38.1.8	Contracted-out employment		
38.1.9	Meaning of contribution, etc.		
38.2	Materials – duties and taxes		
38.3	Fluctuations – work sub-let – Domestic Sub-Contractors		
38.3.1 .2	Provisions relating to Clause 38		
38.4 to 6	Written notice by Contractor		
38.4.1	Timing and effect of written notices		Reasonable time of event; condition precedent to recovery
38.4.2	Agreement – Quantity Surveyor and Contractor		They may agree
38.4.3	Fluctuations added to or deducted from Contract Sum		Not subject to Retention
38.4.4	Evidence and computations by Contractor		To Architect (or QS)
38.4.5	No alteration to Contractor's profit position where Contractor in default over completion		This paid net
38.4.7 and .8			No fluctuations for events after 'scheduled Completion Date'

Ref. No	Subject	Persons primarily concerned	Major contents/comments
38.5	Work, etc. to which clauses 38.1 to 3 are not applicable		Daywork etc.
38.6	Definitions for use with Clause 38		
38.7	Percentage addition to fluctuation payments or allowances		
39	LABOUR AND MATERIALS COST AND TAX FLUCTUATIONS		
39.1.1	Deemed calculation of Contract Sum – rates of wages, etc.		Applicable or promulgated at Date of Tender
39.1.2	Increases or decreases in rates of wages, etc. – payment or allowance		
39.1.3 & .4	Persons employed on site other than 'workpeople'		As Craft operative
39.1.5 & .6	Workpeople – wage-fixing body – reimbursement of fares		
39.2	Contributions, levies and taxes		Note CITB
39.3	Materials, goods, electricity and fuels		Basic list, market price
39.4	Fluctuations – work sub-let – Domestic Sub-Contractors		
39.4.1	Sub-let work – incorporation of provisions to like effect		
39.4.2	Sub-let work – fluctuations – payment to or allowance by Contractor		
39.5 to .7	Provisions relating to Clause 39		
39.5.1	Written notice by Contractor		a.s.a.p. to Architect (or QS)
39.5.2	Timing and effect of written notices		Condition precedent to Contractor's recovery

Ref. No	Subject	Persons primarily concerned	Major contents/comments
39.5.3	Agreement – Quantity Surveyor and Contractor		May agree
39.5.4	Fluctuations added to or deducted from Contract Sum		
39.5.5	Evidence and Computations by Contractor		
39.5.6	No alteration to Contractor's profit		Paid net
39.5.7 and .8	Position where Contractor in default over completion		Not in respect of events after Completion Date
39.6	Work etc. to which clauses 39.1 to 4 not applicable		Daywork, etc.
39.7	Definitions for use with Clause 39		
39.8	Percentage addition to fluctuation payments or allowances		
40 USE OF PRICE ADJUSTMENT FORMULAE			
40.1	Adjustment of Contract Sum – price adjustment formulae for building contracts – Formula Rules		Interim Valuation required for each Interim Certificate Market Price at Date of Tender – use 'manual' system
40.2	Amendment to clause 30 – interim valuations and payments		
40.3	Fluctuations – articles manufactured outside the UK		
40.4	(Number not used)		
40.5	Power to agree – Quantity Surveyor and Contractor		May agree alterations to the methods and procedures
40.6	Position where Monthly Bulletins are delayed, etc.		'Fair' adjustments and use indices retrospectively when available
40.7	Formula adjustment – failure to complete		Over-run – use index at Completion Date

Ref. No	Subject	Persons primarily concerned	Major contents/comments
CONDITIONS: PART 4: SETTLEMENT OF DISPUTES – ADJUDICATION – ARBITRATION – LEGAL PROCEEDINGS			
41A.1	Application	Empl. Contr.	Applies Adjudication to Contract
41A.2	Adjudicator	Empl. Contr.	Agree Adjudicator or nomination
41A.3	Death or incapacity	Empl. Contr. Arch.	Agree new Adjudicator or nomination
41A.4	Referral		Give notice, particulars of difference, summary of contentions, statement of relief
41A.5	Conduct	Adjudicator Empl. Contr.	Detail of the conduct of the Adjudication
41A.6	Fees and expenses	Adjudicator	Decision to deal with fees
41A.7	Effect	All	Decision binding until arbitration litigation or agreement; parties to give effect to decision
41A.8	Immunity	Adjudicator	Adjudicator
Note: 41A given effect due to HGCR			
41B	ADJUDICATION		
41B.1		Empl. Contr.	Arbitrator
41B.2		Empl. Contr. NS/C	Related disputes
41B.3		Empl. Contr.	Joinder
41B.4		Empl. Contr. Arch.	Application to court
41B.5			
41B.6		Art.	CIMAR Arbitration Rules apply
41C		Empl. Contr.	Litigation
CONDITIONS: PART 5: PERFORMANCE SPECIFIED WORK			
42.1	Meaning of Performance Specified Work	Contr. Empl.	Identified in Appendix, on Contract Drawings, in BQ; to be provided by Contractor
42.2	Contractor's Statement	Contr. Empl.	Contractor's Statement for executing the work.
42.3	Contents of Contractor's Statement	Contr. Arch.	
42.4	Time for Contractor's Statement	Contr. Arch.	If no date, at a reasonable time.

Ref. No	Subject	Persons primarily concerned	Major contents/comments
42.5	Architect's notice to amend Contractor's Statement	Contr. Arch.	Remedy for deficiencies; Contractor retains responsibility.
42.6	Architect's notice of deficiency in Contractor's Statement	Arch.	Architect to give notice of any deficiency discovered in Statement.
42.7	Definition of provisional sum for Performance Specified Work.	Contr.	In BQ; sufficient information for Contractor to price.
42.8	Instructions of the Architect on other provisional sums	Arch.	No AI to create Performance Specified Work.
42.9	Preparation of Contract Bills	QS	No
42.10	Provisional sum for Performance Specified Work – errors or omissions in Contract Bills	Contr. Arch. QS	Corrections as Variations.
42.11	Variations in respect of Performance Specified Work	Arch. Contr. QS	
42.12	Agreement for additional Performance Specified Work	Arch. Contr. QS Empl.	No Variation to create Performance Specified Work unless Employer and Contractor so agree. Required if not provided by BQ.
42.13	Analysis	Contr.	
42.14	Integration of Performance Specified Work	Arch. Contr.	AIs may be needed – to integrate with design.
42.15	Compliance with Architect's instructions – Contractor's notice of injurious affection	Contr. Arch.	
42.16	Delay by Contractor in providing the Contractor's statement	Contr. Arch.	No extension of time except for Relevant Events.
42.17	Performance Specified Work – Contractor's obligation	Contr.	As generally but no guarantee of fitness for purpose.
42.18	Nomination excluded	NS/C NSup.	No Performance Specified Work under nominations.
	CODE OF PRACTICE: REFERRED TO IN CLAUSE 8.4.4		
		Arch. Contr.	To assist operation of Clause 8.4.4 Criteria to assist Architect and Contractor to decide appropriate opening up and testing.

Ref. No	Subject	Persons primarily concerned	Major contents/comments
APPENDIX			
ANNEX 1	Bond for advance payment	Empl. Contr.	Advance Payment Bond
	Bond for off-site materials and goods	Empl. Contr.	Materials Off-site Bond
SUPPLEMENTAL PROVISIONS (the VAT Agreement)		Empl. Contr.	
1	Interim payments – addition of VAT		
1.1	Written assessment by Contractor		Amounts subject to positive rate of tax and rate(s) applicable
1.2	Employer to calculate amount of tax due – Employer's right of reasonable objection		
1.3	Written final statement – VAT liability of Contractor – recovery from Employer		
1.4	Contractor to issue receipt as tax invoice		When Contractor receives payment
2.	Value of supply – liquidated damages to be disregarded		
3.	Employer's right to challenge tax claimed by Contractor	Commissioners	Refer to Commissioners?
4.	Discharge of Employer from liability to pay tax to the Contractor		
5.	Awards by Arbitrator or court		
6.	Arbitration provision excluded		
7.	Employer's right where receipt not provided		
ANNEX2	Supplemental provisions for EDI	Empl. Contr.	Electronic Data Interchange

Bibliography/Sources

Hudson's Building and Engineering Contracts (11th Edition) and Supplement, I.N. Duncan Wallace, Sweet and Maxwell, 1995.

Contractors' Guide to the Joint Contracts Tribunal's Standard Forms of Building Contract, 1978, Vincent Powell-Smith, IPC Building and Contract Journals Ltd.

The Standard Forms of Building Contract, Walker-Smith and Close, Charles Knight & Co. Ltd.

JCT Guide to the Standard Form of Building Contract 1980 Edition, Joint Contracts Tribunal, RIBA Publications Ltd.

Introduction to English Law (10th Edition), P.S. James, Butterworths.

John Sims – series of contributions concerning the JCT Standard Form of Building Contract 1980 Edition – from *Building*, commencing 5 February 1980.

Determination of Employment under the Standard Forms of Contract for Construction Works, A.T. Ginnings, *The Quantity Surveyor*, February 1978, pp 97–101.

Building Law Reports, J. Parris (ed), George Godwin.

Construction Law, M.F. James, Macmillan Press, 1994.

Construction Law Reports, V. Powell-Smith, Architectural Press.

V. Powell-Smith – regular articles in Contract Journal.

Sub-Contracting under the JCT Standard Forms of Building Contract, Jennie Price, Macmillan Press, 1994.